Dylan Jones • Mehrdad Tamiz • Jana Ries
Editors

New Developments in Multiple Objective and Goal Programming

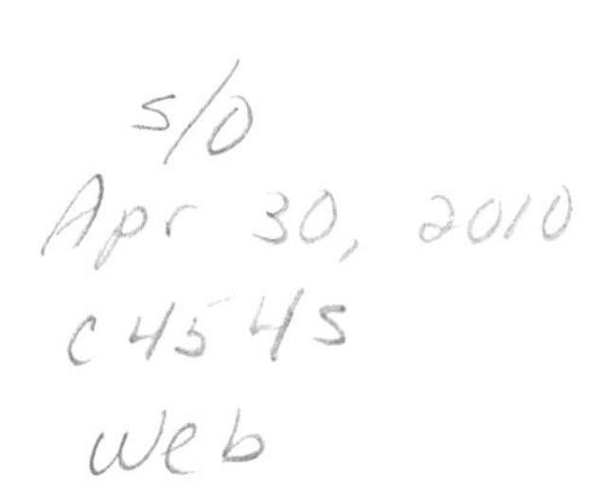

Editors
Dr. Dylan Jones
Dr. Jana Ries
University of Portsmouth
Department Mathematics
Lion Gate Building, Lion Terrace
Portsmouth, Hampshire PO1 3HE
United Kingdom
dylan.jones@port.ac.uk
jana.ries@port.ac.uk

Professor Mehrdad Tamiz
Kuwait University
College of Business and Administration
Department of Quantitative Methods and
Information Systems
P.O. Box 5486
Safat 13055
Kuwait
mehrdad@cba.edu.kw

ISSN 0075-8442
ISBN 978-3-642-10353-7 e-ISBN 978-3-642-10354-4
DOI 10.1007/978-3-642-10354-4
Springer Heidelberg Dordrecht London New York

Library of Congress Control Number: 2010922887

Cover design: SPi Publisher Services

Printed on acid-free paper

Springer is part of Springer Science+Business Media (www.springer.com)

Preface

This volume comprises the proceedings of the 8th International Conference on Multiple Objective and Goal Programming: Theories and Applications (MOPGP08). This conference was held in Portsmouth, United Kingdom on the 24th–26th September 2008. The conference was attended by 59 delegates. The countries represented were Belgium, Finland, France, Germany, Italy, Kuwait, Poland, Saudi Arabia, South Africa, Spain, Switzerland, United Kingdom, and United States.

The MOPGP conference series is now a well-established discussion forum for practitioners and academics working in the field of multiple objective and goal programming. Conferences are held every 2 years, with previous conferences taking place in Portsmouth, UK in 1994; in Malaga, Spain in 1996; in Quebec City, Canada in 1998; in Ustron, Poland in 2000; in Nara, Japan in 2002; in Hammamet, Tunisia in 2004; and in Tours, France in 2006.

The selection of papers published in this volume reflects the breadth of the techniques and applications in the field of multiple objective and goal programming. Applications from the fields of supply chain management, financial portfolio selection, financial risk management, insurance, medical imaging, sustainability, nurse scheduling, project management, and the interface with data envelopment analysis give a good reflection of current usage. A pleasing variety of techniques are used, including models with fuzzy, group-decision, stochastic, interactive, and binary aspects. Additionally, two papers from the upcoming area of multi-objective evolutionary algorithms are included.

As organisers of the MOGP08 conference, we thank all the participants and presenters at the conference for helping the MOPGP series to continue to define and develop the state-of-the-art in the field of multiple objective and goal programming. We also thank the University of Portsmouth and their conference management team for their support and help in ensuring a smooth running conference. We also thank the members of the local organising committee – Rania Azmi, Patrick Beullens, Alessio Ishizaka, Ashraf Labib, Ming Yip Lam, Kevin Willis, and Nerda Zaibidi from the University of Portsmouth and Ersilia Liguigli from the Politecnico di Torino – for their hard work and support before and during the conference.

Finally, we thank the referees for their efficiency in producing reports during the review process. We hope that this volume will inform the reader in the start-of-the-art of both theory and application in the area of multiple objective and goal programming.

Portsmouth, UK and Kuwait City, Kuwait
August 2009

Dylan Jones
Mehrdad Tamiz
Jana Ries

Contents

List of Contributors

Abbas Afshar Department of Civil Engineering, Iran University of Science and Technology, Narmak, Tehran 16844, Iran, a_afshar@iust.ac.ir

Amir Azaron Institut für Fördertechnik und Logistiksysteme, Universität Karlsruhe (TH), Gotthard-Franz-Str. 8, 76137 Karlsruhe, Germany
and
Department of Financial Engineering and Engineering Management, School of Science and Engineering, Reykjavik University, Reykjavik, Iceland, amir.azaron@ucd.ir

Rania Azmi University of Portsmouth, UK, rania.azmi@port.ac.uk
and
P.O. Box 1529, Salmiya, 22016, Kuwait, Rania.A.Azmi@gmail.com

Johannes Bader Computer Engineering and Networks Lab (TIK), Institute TIK, ETH Zurich, Gloriastr. 35, 8092 Zurich, Switzerland, Johannes.Bader@tik.ee.ethz.c

Alejandro Balbás University Carlos III of Madrid, CL. Madrid, 126, 28903 Getafe, Madrid, Spain, alejandro.balbas@uc3m.es

Beatriz Balbás University Rey Juan Carlos, Paseo Artilleros s/n, 28032-Vicálvaro, Madrid, Spain, beatriz.balbas@urjc.es

Raquel Balbás Department of Actuarial and Financial Economics, University Complutense of Madrid, 28223 Pozuelo de Alarcón, Madrid, Spain, raquel.balbas@ccee.ucm.es

Enrique Ballestero Escuela Politécnica Superior de Alcoy, 03801 Alcoy, Alicante, Spain, eballe@esp.upv.es

Dimo Brockhoff Computer Engineering and Networks Lab, ETH Zurich, 8092 Zurich, Switzerland, dimo.brockhoff@tik.ee.ethz.ch

Luis Diaz-Balteiro ETS Ingenieros de Montes, Ciudad Universitaria s/n, 28040 Madrid, Spain, luis.diaz.balteiro(at)upm.es

Kai Furmans Institut für Fördertechnik und Logistiksysteme, Universität Karlsruhe (TH), Karlsruhe, Germany, kai.furmans@ifl.uni-karisruhe.de

Kaisa Miettinen Department on Mathematical Information Technology, University of Jyvaskyla, P.O. Box 35, University of Jyvaskyla, 40014, Finland, Kaisa.Miettinen@jyu.fi

Mohammad Modarres Department of Industrial Engineering, Sharif University of Technology, Tehran, Iran, Modarres@sharif.edu

Maciej Nowak Department of Operations Research, The Karol Adamiecki University of Economics, ul. Bogucicka 14, 40-226 Katowice, Poland, nomac@ae.katowice.pl,Maciej.Nowak@ae.katowice.pl

Jan-Erik Palmgren Department of Oncology, Kuopio University Hospital, P.O. Box 1777, 70211 Kuopio, Finland, Jan-Erik.Palmgren@kuh.fi

Carlos Romero Research Group "Economics for a Sustainable Environment", Technical University of Madrid, Madrid, Spain, carlos.romero@upm.es

Henri Ruotsalainen Department of Physics, University of Kuopio, P.O. Box 1627, 70211 Kuopio, Finland, Henri.Ruotsalainen@uku.fi

Mehrdad Tamiz College of Business Administration, Kuwait University, Kuwait, mehrdad@cba.edu.kw

Chris Tofallis The Business School, University of Hertfordshire, College Lane, Hatfield, Hertfordshire AL10 9AB, UK, c.tofallis@herts.ac.uk

Roberto Voces Technical University of Madrid, Ciudad Universitaria s/n, 28040 Madrid, Spain, vocesr@gmail.com

Mohammed Ali Yaghoobi Department of Statistics, Faculty of Mathematics and Computer, Shahid Bahonar University of Kerman, Kerman 76169-14111, Iran, yaghoobi@mail.uk.ac.ir

Eckart Zitzler Computer Engineering and Networks Lab, ETH Zurich, 8092 Zurich, Switzerland, eckart.zitzler@tik.ee.ethz.ch

Multi-Objective Stochastic Programming Approaches for Supply Chain Management

Amir Azaron, Kai Furmans, and Mohammad Modarres

Abstract A multi-objective stochastic programming model is developed to design robust supply chain configuration networks. Demands, supplies, processing, and transportation costs are all considered as the uncertain parameters, which will be revealed after building the sites at the strategic level. The decisions about the optimal flows are made at the tactical level depending upon the actual values of uncertain parameters. It is also assumed that the suppliers are unreliable. To develop a robust model, two additional objective functions are added into the traditional supply chain design problem. So, the proposed model accounts for the minimization of the expected total cost and the risk, reflected by the variance of the total cost and the downside risk or the risk of loss. Finally, different simple and interactive multi-objective techniques such as goal attainment, surrogate worth trade-off (SWT), and STEM methods are used to solve the proposed multi-objective model.

1 Introduction

A supply chain (SC) is a network of suppliers, manufacturing plants, warehouses, and distribution channels organized to acquire raw materials, convert these raw materials to finished products, and distribute these products to customers. The concept of supply chain management (SCM), which appeared in the early 1990s, has recently raised a lot of interest since the opportunity of an integrated management of the supply chain can reduce the propagation of unexpected/undesirable events through the network and can affect decisively the profitability of all the members.

A crucial component of the planning activities of a manufacturing firm is the efficient design and operation of its supply chain. Strategic level supply chain planning

A. Azaron (✉)
Institut für Fördertechnik und Logistiksysteme, Universität Karlsruhe (TH), Karlsruhe, Germany
and
Department of Financial Engineering and Engineering Management, School of Science and Engineering, Reykjavik University, Reykjavik, Iceland
e-mail: amir.azaron@ucd.ir

D. Jones et al. (eds.), *New Developments in Multiple Objective and Goal Programming*, Lecture Notes in Economics and Mathematical Systems 638,
DOI 10.1007/978-3-642-10354-4_1,

involves deciding the configuration of the network, i.e., the number, location, capacity, and technology of the facilities. The tactical level planning of supply chain operations involves deciding the aggregate quantities and material flows for purchasing, processing, and distribution of products. The strategic configuration of the supply chain is a key factor influencing efficient tactical operations, and therefore has a long lasting impact on the firm. Furthermore, the fact that the supply chain configuration involves the commitment of substantial capital resources over long periods of time makes the supply chain design problem an extremely important one.

Many attempts have been made to model and optimize supply chain design, most of which are based on deterministic approaches, see for example Bok et al. (2000), Timpe and Kallrath (2000), Gjerdrum et al. (2000), and many others. However, most real supply chain design problems are characterized by numerous sources of technical and commercial uncertainty, and so the assumption that all model parameters, such as cost coefficients, supplies, demands, etc., are known with certainty is not realistic.

In order to take into account the effects of the uncertainty in the production scenario, a two-stage stochastic model is proposed in this paper. Decision variables which characterize the network configuration, namely those binary variables which represent the existence and the location of plants and warehouses of the supply chain are considered as first-stage variables – it is assumed that they have to be taken at the strategic level before the realization of the uncertainty. On the other hand, decision variables related to the amount of products to be produced and stored in the nodes of the supply chain and the flows of materials transported among the entities of the network are considered as second-stage variables, corresponding to decisions taken at the tactical level after the uncertain parameters have been revealed.

There are a few research works addressing comprehensive (strategic and tactical issues simultaneously) design of supply chain networks using two-stage stochastic models. MirHassani et al. (2000) considered a two-stage model for multi-period capacity planning of supply chain networks, and used Benders decomposition to solve the resulting stochastic integer program. Tsiakis et al. (2001) considered a two-stage stochastic programming model for supply chain network design under demand uncertainty, and developed a large-scale mixed-integer linear programming model for this problem. Alonso-Ayuso et al. (2003) proposed a branch-and-fix heuristic for solving two-stage stochastic supply chain design problems. Santoso et al. (2005) integrated a sampling strategy with an accelerated Benders decomposition to solve supply chain design problems with continuous distributions for the uncertain parameters. However, the robustness of decision to uncertain parameters is not considered in above studies.

Azaron et al. (2008) developed a multi-objective stochastic programming approach for designing robust supply chains. The objective functions of this model are (1) the minimization of the sum of current investment costs and the expected future processing, transportation, shortage, and capacity expansion costs, (2) the minimization of the variance of the total cost, and (3) the minimization of the financial risk or the probability of not meeting a certain budget. Then, they used goal attainment technique, see Hwang and Masud (1979) for details, to solve the resulting multi-objective problem.

This method has the same disadvantages as those of goal programming; namely, the preferred solution is sensitive to the goal vector and the weighting vector given by the decision maker. To overcome this drawback, we also use interactive multi-objective techniques with explicit or implicit trade-off information given such as SWT and STEM methods, see Hwang and Masud (1979) for details, to solve the problem. The other advantage of this paper over (Azaron et al. 2008) is that we minimize downside risk or risk of loss instead of financial risk. By applying this concept, we avoid the use of binary variables to determine the financial risk, which significantly reduces the computational time to solve the final large scale mixed-integer nonlinear programming problem.

To the best of our knowledge, only ε-constraint method (Guillen et al. 2005), fuzzy optimization (Chen and Lee 2004), and goal attainment method (Azaron et al. 2008) have been used to solve existing multi-objective supply chain design models. In this paper, we use interactive multi-objective techniques to solve the problem.

The paper is organized as follows. In Sect. 2, we describe the supply chain design problem. In Sect. 3, we explain the details of multi-objective techniques to solve the problem. Section 4 presents the computational experiments. Finally, we draw the conclusion of the paper in Sect. 5.

2 Problem Description

We first describe a deterministic mathematical formulation for the supply chain design problem. Consider a supply chain network $G = (N, A)$, where N is the set of nodes and A is the set of arcs. The set N consists of the set of suppliers S, the set of possible processing facilities P, and the set of customer centers C, i.e., $N = S \cup P \cup C$. The processing facilities include manufacturing centers M and warehouses W, i.e., $P = M \cup W$. Let K be the set of products flowing through the supply chain.

The supply chain configuration decisions consist of deciding which of the processing centers to build. We associate a binary variable y_i to these decisions: $y_i = 1$ if processing facility i is built, and 0 otherwise. The tactical decisions consist of routing the flow of each product $k \in K$ from the suppliers to the customers. We let x_{ij}^k denote the flow of product k from a node i to a node j of the network where $(ij) \in A$, and z_j^k denote shortfall of product k at customer centre j, when it is impossible to meet demand. A deterministic mathematical model for this supply chain design problem is formulated as follows (see Santoso et al. (2005) for more details):

$$\text{Min} \sum_{i \in P} c_i y_i + \sum_{k \in K} \sum_{(ij) \in A} q_{ij}^k x_{ij}^k + \sum_{k \in K} \sum_{j \in C} h_j^k z_j^k \tag{1a}$$

s.t.

$$y \in Y \subseteq \{0,1\}^{|P|} \tag{1b}$$

$$\sum_{i \in N} x_{ij}^k - \sum_{l \in N} x_{jl}^k = 0 \qquad \forall j \in P, \quad \forall k \in K \tag{1c}$$

$$\sum_{i \in N} x_{ij}^k + z_j^k \geq d_j^k \qquad \forall j \in C, \quad \forall k \in K \tag{1d}$$

$$\sum_{j \in N} x_{ij}^k \leq s_i^k \qquad \forall i \in S, \quad \forall k \in K \tag{1e}$$

$$\sum_{k \in K} r_j^k \left(\sum_{i \in N} x_{ij}^k \right) \leq m_j y_j \qquad \forall j \in P \tag{1f}$$

$$x_{ij}^k \geq 0 \qquad \forall (ij) \in A, \quad \forall k \in K \tag{1g}$$

$$z_j^k \geq 0 \qquad \forall j \in C, \quad \forall k \in K \tag{1h}$$

The objective function (1a) consists of minimizing the total investment, production/transportation, and shortage costs. Constraint (1b) enforces the binary nature of the configuration decisions for the processing facilities. Constraint (1c) enforces the flow conservation of product k across each processing node j. Constraint (1d) requires that the total flow of product k to a customer node j plus shortfall should exceed the demand d_j^k at that node. Constraint (1e) requires that the total flow of product k from a supplier node i should be less than the supply s_i^k at that node. Constraint (1f) enforces capacity constraints of the processing nodes. Here, r_j^k and m_j denote per-unit processing requirement for product k at node j and capacity of facility j, respectively.

We now propose a stochastic programming approach based on a recourse model with two stages to incorporate the uncertainty associated with demands, supplies, processing/transportation, shortage, and capacity expansion costs. It is also assumed that we have the option of expanding the capacities of sites after the realization of uncertain parameters. Considering $\xi = (d, s, q, h, f)$ as the corresponding random vector, the two-stage stochastic model, in Matrix form, is formulated as follows (see Azaron et al. (2008) for details):

$$\text{Min } c^T y + E[G(y, \xi)] \quad \text{[Expected Total Cost]} \tag{2a}$$

s.t.

$$y \in Y \subseteq \{0,1\}^{|P|} \quad \text{[Binary Variables]} \tag{2b}$$

where $G(y, \xi)$ is the optimal value of the following problem:

$$\text{Min } q^T x + h^T z + f^T e \tag{2c}$$

s.t.

$$Bx = 0 \quad \text{[Flow Conservation]} \tag{2d}$$
$$Dx + z \geq d \quad \text{[Meeting Demand]} \tag{2e}$$
$$Sx \leq s \quad \text{[Supply Limit]} \tag{2f}$$
$$Rx \leq My + e \quad \text{[Capacity Constraint]} \tag{2g}$$
$$e \leq Oy \quad \text{[Capacity Expansion Limit]} \tag{2h}$$
$$x \in R_+^{|A|\times|K|},\ z \in R_+^{|C|\times|K|}\ e \in R_+^{|P|} \quad \text{[Continuous Variables]} \tag{2i}$$

Above vectors c, q, h, f, d, and s correspond to investment costs, processing/transportation costs, shortfall costs, expansion costs, demands, and supplies, respectively. The matrices B, D, and S are appropriate matrices corresponding to the summations on the left-hand-side of the expressions (1c)–(1e), respectively. The notation R corresponds to a matrix of r_j^k, and the notation M corresponds to a matrix with m_j along the diagonal. e and O correspond to capacity expansions and expansion limits, respectively.

Note that the optimal value $G(y, \xi)$ of the second-stage problem (2c)–(2i) is a function of the first stage decision variable y and a realization $\xi = (d, s, q, h, f)$ of the uncertain parameters. The expectation in (2a) is taken with respect to the joint probability distribution of uncertain parameters.

In this paper, the uncertainty is represented by a set of discrete scenarios with given probability of occurrence. It is also assumed that suppliers are unreliable and their reliabilities are known in advance. The role of unreliable suppliers is implicitly considered in the model by properly way of generating scenarios. It means that in case of having an unreliable supplier, its supply value is set to zero in the corresponding scenarios, see Azaron et al. (2008) for more details.

3 Multi-Objective Techniques

As explained, to develop a robust model, two additional objective functions are added into the traditional supply chain design problem. The first is the minimization of the variance of the total cost, and the second is the minimization of the downside risk or the risk of loss. The definition of downside risk or the expected total loss is:

$$DRisk = \sum_{l=1}^{L} p_l Max(Cost_l - \Omega, 0) \tag{3}$$

where p_l, Ω, and $Cost_l$ represent the occurrence probability of the lth scenario, available budget, and total cost when the lth scenario is realized, respectively. The downside risk can be calculated as follows:

$$
\begin{aligned}
&DRisk = \sum_{l=1}^{L} p_l DR_l \\
&DR_l \geq Cost_l - \Omega \qquad \forall l \\
&DR_l \geq 0 \qquad \forall l
\end{aligned} \tag{4}
$$

The proper multi-objective stochastic model for our supply chain design problem will be:

$$\text{Min } f_1(x) = c^T y + \sum_{l=1}^{L} p_l \left(q_l^T x_l + h_l^T z_l + f_l^T e_l \right) \text{[Expected Total Cost]} \tag{5a}$$

$$\text{Min } f_2(x) = \sum_{l=1}^{L} p_l \left(q_l^T x_l + h_l^T z_l + f_l^T e_l - \sum_{l=1}^{L} p_l \left(q_l^T x_l + h_l^T z_l + f_l^T e_l \right) \right)^2 \text{[Variance]} \tag{5b}$$

$$\text{Min } f_3(x) = \sum_{l=1}^{L} p_l DR_l \quad \text{[Downside Risk]} \tag{5c}$$

s.t.

$$Bx_l = 0 \quad l = 1, \ldots, L \tag{5d}$$

$$Dx_l + z_l \geq d_l \quad l = 1, \ldots, L \tag{5e}$$

$$Sx_l \leq s_l \quad l = 1, \ldots, L \tag{5f}$$

$$Rx_l \leq My + e_l \quad l = 1, \ldots, L \tag{5g}$$

$$e_l \leq Oy \quad l = 1, \ldots, L \tag{5h}$$

$$c^T y + q_l^T x_l + h_l^T z_l + f_l^T e_l - \Omega \leq DR_l \quad l = 1, \ldots, L \tag{5i}$$

$$y \in Y \subseteq \{0, 1\}^{|P|} \tag{5j}$$

$$x \in R_+^{|A| \times |K| \times L}, \quad z \in R_+^{|C| \times |K| \times L}, \quad e \in R_+^{|P| \times L}, \quad DR \in R_+^L \tag{5k}$$

3.1 Goal Attainment Technique

Goal attainment method is one of the multi-objective techniques with priori articulation of preference information given. In this method, the preferred solution is sensitive to the goal vector and the weighting vector given by the decision maker; the same as goal programming technique. However, goal attainment method has fewer variables to work with and is a one-stage method, unlike interactive multi-objective techniques, so it will be computationally faster.

This method requires setting up a goal and weight, b_j and $g_j (g_j \geq 0)$ for $j = 1, 2, 3$, for the three mentioned objective functions. The g_j relates the relative under-attainment of the b_j. For under-attainment of the goals, a smaller g_j is associated with the more important objectives. When g_j approaches 0, then the associated objectivefunction should be fully satisfied or the corresponding objective function value should be less than or equal its goal b_j. g_j, $j = 1, 2, 3$, are generally normalized so that $\sum_{j=1}^{3} g_j = 1$. The proper goal attainment formulation for our problem is:

$$\text{Min } w \tag{6a}$$

s.t.

$$c^T y + \sum_{l=1}^{L} p_l \left(q_l^T x_l + h_l^T z_l + f_l^T e_l \right) - g_1 w \leq b_1 \tag{6b}$$

$$\sum_{l=1}^{L} p_l \left(q_l^T x_l + h_l^T z_l + f_l^T e_l - \sum_{l=1}^{L} p_l \left(q_l^T x_l + h_l^T z_l + f_l^T e_l \right) \right)^2$$
$$-g_2 w \leq b_2 \tag{6c}$$

$$\sum_{l=1}^{L} p_l DR_l - g_3 w \leq b_3 \tag{6d}$$

$$Bx_l = 0 \quad l = 1, \ldots, L \tag{6e}$$

$$Dx_l + z_l \geq d_l \quad l = 1, \ldots, L \tag{6f}$$

$$Sx_l \leq s_l \quad l = 1, \ldots, L \tag{6g}$$

$$Rx_l \leq My + e_l \quad l = 1, \ldots, L \tag{6h}$$

$$e_l \leq Oy \quad l = 1, \ldots, L \tag{6i}$$

$$c^T y + q_l^T x_l + h_l^T z_l + f_l^T e_l - \Omega \leq DR_l \quad l = 1, \ldots, L \tag{6j}$$

$$y \in Y \subseteq \{0, 1\}^{|P|} \tag{6k}$$

$$x \in R_+^{|A| \times |K| \times L}, \quad z \in R_+^{|C| \times |K| \times L}, \quad e \in R_+^{|P| \times L}, \quad DR \in R_+^L \tag{6l}$$

Lemma 1. *If* (y^*, x^*, z^*, e^*) *is Pareto-optimal, then there exists a b and g pair such that* (y^*, x^*, z^*, e^*) *is an optimal solution to the optimization problem (6).*

The optimal solution using this formulation is sensitive to b and g. Depending upon the values for b, it is possible that g does not appreciably influence the optimal solution. Instead, the optimal solution can be determined by the nearest Pareto-optimal solution from b. This might require that g be varied parametrically to generate a set of Pareto-optimal solutions.

3.2 STEM Method

The main drawback of the goal attainment technique to solve ***(5) is that the prefered solution extremely depends on the goals and weights. To overcome this drawback, we resort to STEM and SWT methods, which are two main interactive multi-objective techniques, to solve the multi-objective model.

In this subsection, we explain the details of the STEM method, which is an interactive approach with implicit trade-off information given. STEM allows the decision maker (DM) to learn to recognize good solutions and the relative importance of the objectives. In this method, phases of computation alternate (interactively) with phases of decision. The major steps of the STEM method to solve the multi-objective problem are:

Step 0. Construction of a pay-off table:
A pay-off table is constructed before the first interactive cycle. Let f_j^*, $j = 1, 2, 3$, be feasible ideal solutions of the following three problems:

Min $f_j(x), \quad j = 1, 2, 3$

s.t.

$$x \in S \text{ (feasible region of (5))} \tag{7}$$

In the pay-off table, row j corresponds to the solution vector x^*, which optimizes the objective function f_j. A z_{ij} is the value taken on by the ith objective f_i when the jth objective f_j reaches its optimum f_j^*.
Step 1. Calculation phase:
At the mth cycle, the feasible solution to the problem ***(8) is sought, which is the "nearest", in the MINIMAX sense, to the ideal solution f_j^*:

Min γ

s.t.

$$\begin{aligned} &\gamma \geq (f_j(x) - f_j^*)\pi_j, \quad j = 1, 2, 3 \\ &x \in X^m \\ &\gamma \geq 0 \end{aligned} \tag{8}$$

where X^m includes S plus any constraint added in the previous $(m - 1)$ cycles; π_j gives the relative importance of the distances to the optima. Let us consider the jth column of the pay-off table. Let $f_j^{\max}$ and $f_j^{\min}$ be the maximum and minimum values; then π_j, $j = 1, 2, 3$, are chosen such that $\pi_j = \frac{\alpha_j}{\sum_i \alpha_i}$, where $\alpha_j = \frac{f_j^{\max} - f_j^{\min}}{f_j^{\max}}$.

From the above equations, we conclude that if the value of f_j does not vary much from the optimum solution by varying x, the corresponding objective is not sensitive to a variation in the weighting values; therefore, a small weight π_j can be assigned to this objective function.
Step 2. Decision phase:
The compromise solution x^m is presented to the DM. If some of the objectives are satisfactory and others are not, the DM relaxes a satisfactory objective f_j^m enough to allow an improvement of the unsatisfactory objectives in the next iterative cycle.

The DM gives Δf_j as the amount of acceptable relaxation. Then, for the next cycle the feasible region is modified as:

$$X^{m+1} = \begin{cases} X^m \\ f_j(x) \leq f_j(x^m) + \Delta f_j, if \quad j = 1,2,3 \\ f_i(x) \leq f_i(x^m), if \quad i = 1,2,3, i \neq j \end{cases} \tag{9}$$

The weight π_j is set to zero and the calculation phase of cycle $m+1$ begins.

3.3 *Surrogate Worth Trade-Off (SWT) Method*

In this subsection, we explain the details of the SWT method, which is an interactive approach with explicit trade-off information given. It is a virtue that all the alternatives during the solution process are non-dominated. Thus the decision maker is not bothered with any other kind of solutions. The major steps in the SWT method to solve the multi-objective problem (5) are:

Step 1. Determine the ideal solution for each of the objectives in problem (5). Then set up the multi-objective problem in the form of (10).

Min $f_1(x)$

s.t.

$$\begin{aligned} & f_2(x) \leq \varepsilon_2 \\ & f_3(x) \leq \varepsilon_3 \\ & x \in S \text{ (Feasible region of problem (5))} \end{aligned} \tag{10}$$

Step 2. Identify and generate a set of non-dominated solutions by varying ε 's parametrically in problem (10). Assuming μ_j, $j = 2, 3$, as the Lagrange multipliers corresponding with the first set of constraints of problem (10), the non-dominated solutions are the ones, which have non-zero values for μ_j.
Step 3. Interact with the DM to assess the surrogate worth function w_j, or the DM's assessment of how much (from -10 to 10) he prefers trading μ_j marginal units of the first objective for one marginal unit of the jth objective $f_j(x)$, given the other objectives remaining at their current values.
Step 4. Isolate the indifference solutions. The solutions, which have $w_j = 0$ for all j, are said to be indifference solutions. If there exists no indifference solution, develop approximate relations for all worth functions $w_j = \widehat{w}_j(f_j, j = 2, 3)$, by multiple regressions. Solve the simultaneous equations $\widehat{w}_j(f) = 0$ for all j to obtain f^* (f^* does not include $f_1{}^*$). Then, solve problem (11). Present this solution to the DM, and ask if this is an indifference solution. If yes, it is a preferred solution; proceed to Step 5. Otherwise, repeat the process of generating more

non-dominated solutions around $\widehat{w}_j(f) = 0$ and refining the estimated f^* until it results in an indifference solution.

Min $f_1(x)$

s.t.

$$\begin{aligned} f_2(x) &\leq f_2^*(x) \\ f_3(x) &\leq f_3^*(x) \\ x &\in S \end{aligned} \tag{11}$$

Step 5. The optimal solution $f_1{}^*$ along with f^* and x^* would be the optimal solution to the multi-objective problem (5).

4 Numerical Experiments

Consider the supply chain network design problem depicted in Fig. 1. A wine company is willing to design its supply chain. This company owns three customer centers located in three different cities L, M, and N, respectively. Uniform-quality wine in bulk (raw material) is supplied from four wineries located in A, B, C, and D. There are four possible locations E, F, G, and H for building the bottling plants.

For simplicity, without considering other market behaviors (e.g., novel promotion, marketing strategies of competitors, and market-share effect in different markets), each market demand merely depends on the local economic conditions. Assume that the future economy is either boom, good, fair, or poor, i.e., four situations with associated probabilities of 0.13, 0.25, 0.45, or 0.17, respectively. The unit production costs and market demands under each scenario are shown in Table 1.

The supplies, transportation costs, and shortage costs are considered as deterministic parameters. (475,000, 425,000, 500,000, 450,000) are investment costs for

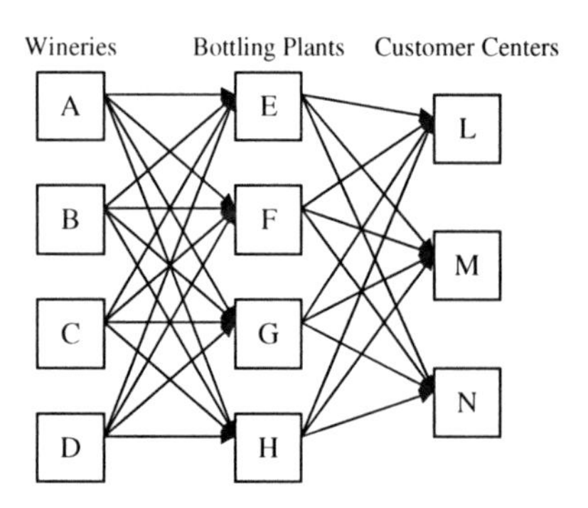

Fig. 1 The supply chain design problem of the wine company

Table 1 Characteristics of the problem

Future economy	Demands			Unit production costs				Probabilities
	L	M	N	E	F	G	H	
Boom	400	188	200	755	650	700	800	0.13
Good	350	161	185	700	600	650	750	0.25
Fair	280	150	160	675	580	620	720	0.45
Poor	240	143	130	650	570	600	700	0.17

Table 2 Pay-off table

Mean	1856986	307077100000	119113
Var	6165288	0	3985288
DRisk	2179694	1495374000	4467.3

building each bottling plant E, F, G, and H, respectively. (65.6, 155.5, 64.3, 175.3, 62, 150.5, 59.1, 175.2, 84, 174.5, 87.5, 208.9, 110.5, 100.5, 109, 97.8) are the unit costs of transporting bulk wine from each winery A, B, C, and D to each bottling plant E, F, G, and H, respectively. (200.5, 300.5, 699.5, 693, 533, 362, 163.8, 307, 594.8, 625, 613.6, 335.5) are the unit costs of transporting bottled wine from each bottling plant E, F, G, and H to each distribution center L, M, and N, respectively. (10,000, 13,000, 12,000) are the unit shortage costs at each distribution center L, M, and N, respectively. (375, 187, 250, 150) are the maximum amount of bulk wine that can be shipped from each winery A, B, C, and D, respectively, if it is reliable. (315, 260, 340, 280) are the capacities of each bottling plant E, F, G, and H, respectively, if it is built.

We also have the option of expanding the capacity of bottling plant F, if it is built. (100, 80, 60, 50) are the unit capacity expansion costs, when the future economy is boom, good, fair or poor, respectively. In addition, we cannot expand the capacity of this plant more than 40 units in any situation. Moreover, winery D is an unreliable supplier and may lose its ability to supply the bottling plants. The reliability of this winery is estimated as 0.9. So, the total number of scenarios for this supply chain design problem is equal to $4 \times 2 = 8$.

The problem attempts to minimize the expected total cost, the variance of the total cost and the downside risk in a multi-objective scheme while making the following determinations:

(a) Which of the bottling plants to build (first-stage variables)?
(b) Flows of materials transported among the entities of the network (second-stage variables)?

First, goal attainment technique is used to solve this multi-objective supply chain design problem (refer to Azaron et al. (2008) to see some results). Then, we use STEM and SWT methods to solve the problem.

In the beginning, we construct the pay-off table, using STEM method, which is shown in Table 2.

Then, we go to the calculation phase and solve the problem (8) using LINGO 10 on a PC Pentium IV 2.1-GHz processor. The compromise solution for the location (strategic) variables is $[1, 1, 1, 0]$. Then, we go to the decision phase and the compromise solution is presented to the DM, who compares its objective vector $f^1 = \left(f_1{}^1, f_2{}^1, f_3{}^1\right) = (2219887, 253346, 39887)$ with the ideal one, $f^* = \left(f_1^*, f_2^*, f_3^*\right) = (1856986, 0, 4467.3)$. If $f_2{}^1$ is satisfactory, but the other objectives are not, the DM must relax the satisfactory objective $f_2{}^1$ enough to allow an improvement of the unsatisfactory objectives in the first cycle. Then, $\Delta f_2 = 999746654$ is considered as the acceptable amount of relaxation, and the feasible region is modified as following in the next iteration cycle:

$$X^2 = \begin{cases} X^1 \\ f_1(x) \leq 2219887 \\ f_2(x) \leq f_2(x^1) + 999746654 = 1000000000 \\ f_3(x) \leq 39887 \end{cases} \tag{12}$$

In this case, three times of the acceptable level of the standard deviation is almost equal to 1,00,000, and comparing this with the expected total cost implies an acceptable amount of relaxation for the variance. In the second cycle, the compromise solution for the location variables is still $[1, 1, 1, 0]$. This compromise solution is again presented to the DM, who compares its objective vector $f^2 = \left(f_1{}^2, f_2{}^2, f_3{}^2\right) = (2132615, 1000000000, 4467.3)$ with the ideal one. If all objectives of the vector f^2 are satisfactory, f^2 is the final solution and the optimal vector including the strategic and tactical variables would be x^2.

The total computational time to solve the problem using STEM method is equal to 18:47 (mm:ss), comparing to 02:26:37 (hh:mm:ss) in generating 55 Pareto-optimal solutions using goal attainment technique, see Azaron et al. (2008) for details.

We also use the SWT method to solve the same problem. Using this method, the single objective optimization problem for generating a set of non-dominated solutions is formulated according to (10). Then, $\varepsilon_j, j = 2, 3$, are varied to obtain several non-dominated solutions.

For example, by considering $\varepsilon_2 = 100000$ and $\varepsilon_3 = 1000$, we obtain a non-dominated solution for the location variables as $[1, 1, 1, 0]$, which can also be considered as an indifference solution by the DM. This indifference solution has the same structure as the STEM compromise solutions, but certainly with different second-stage variables. The corresponding objective vector is $f^* = (f_1{}^*, f_2{}^*, f_3{}^*) = (2143678, 748031500, 4467.3)$. The computational time to get this solution is equal to 06:39 (mm:ss).

Comparing this solution with the final STEM solution shows that however the risk has been reduced in the SWT method, but the expected total cost has been increased, and none of these solutions can dominate the other one.

5 Conclusion

Determining the optimal supply chain configuration is a difficult problem since a lot of factors and objectives must be taken into account when designing the network under uncertainty. The proposed model in this paper accounts for the minimization of the expected total cost, the variance of the total cost, and the downside risk in a multi-objective scheme to design a robust supply chain network. By using this methodology, the trade-off between the expected total cost and risk terms can be obtained. The interaction between the design objectives has also been shown. Therefore, this approach seems to be a good way of capturing the high complexity of the problem.

We used goal attainment, which is a simple multi-objective technique, and STEM and SWT methods, which are two main interactive multi-objective techniques, to solve the problem. The main advantage of these interactive techniques is that the prefered solution does not depend on the goal and weight vectors, unlike traditional goal programming technique. We also avoided using several more binary variables in defining financial risk by introducing downside risk in this paper, which significantly reduced the computational times.

Since the final mathematical model is a large-scale mixed-integer nonlinear program, developing a meta-heuristic approach such as genetic algorithm or simulated annealing will be helpful in terms of computational time.

In case the random data vector follows a known continuous joint distribution, one should resort to a sampling procedure, for example Santoso et al. (2005), to solve the problem. In this case, an integration of sampling strategy along with Benders decomposition technique would be suitable to solve the resulting stochastic mixed-integer program.

Acknowledgements This research is supported by Alexander von Humboldt-Stiftung and Iran National Science Foundation (INSF).

References

Alonso-Ayuso A, Escudero LF, Garin A, Ortuno MT, Perez G (2003) An approach for strategic supply chain planning under uncertainty based on stochastic 0–1 programming. J Global Optim 26:97–124

Azaron A, Brown KN, Tarim SA, Modarres M (2008) A multi-objective stochastic programming approach for supply chain design considering risk. Int J Prod Econ 116:129–138

Bok JK, Grossmann IE, Park S (2000) Supply chain optimization in continuous flexible process networks. Ind Eng Chem Res 39:1279–1290

Chen CL, Lee WC (2004) Multi-objective optimization of multi-echelon supply chain networks with uncertain demands and prices. Comput Chem Eng 28:1131–1144

Gjerdrum J, Shah N, Papageorgiou LG (2000) A combined optimisation and agent-based approach for supply chain modelling and performance assessment. Prod Plann Control 12:81–88

Guillen G, Mele FD, Bagajewicz MJ, Espuna A, Puigjaner L (2005) Multiobjective supply chain design under uncertainty. Chem Eng Sci 60:1535–1553

Hwang CL, Masud ASM (1979) Multiple objective decision making. Springer, Berlin
MirHassani SA, Lucas C, Mitra G, Messina E, Poojari CA (2000) Computational solution of capacity planning models under uncertainty. Parallel Comput 26:511–538
Santoso T, Ahmed S, Goetschalckx M, Shapiro A (2005) A stochastic programming approach for supply chain network design under uncertainty. Eur J Oper Res 167:96–115
Timpe CH, Kallrath J (2000) Optimal planning in large multi-site production networks. Eur J Oper Res 126:422–435
Tsiakis P, Shah N, Pantelides CC (2001) Design of multiechelon supply chain networks under demand uncertainty. Ind Eng Chem Res 40:3585–3604

A Review of Goal Programming for Portfolio Selection

Rania Azmi and Mehrdad Tamiz

Abstract Goal Programming (GP) is the most widely used approach in the field of multiple criteria decision making that enables the decision maker to incorporate numerous variations of constraints and goals, particularly in the field of Portfolio Selection (PS). This paper gives a brief review of the application of GP and its variants to Portfolio Selection and analysis problems. The paper firstly discusses the Multi-Criteria Decision Analysis in PS context in which GP is introduced as an important approach to PS Problems. An overview of performance measurement in portfolio selection context is also provided. Amongst the concluding remarks many issues in PS that may be addressed by GP such as multi-period, different measures of risk, and extended factors influencing portfolio selection are listed.

1 Introduction

Finance theory has produced a variety of models that attempt to provide some insight into the environment in which financial decisions are made. By definition, every model is a simplification of reality. Hence, even if the data fail to reject the model, the decision maker may not necessarily want to use the model as a dogma. At the same time, the notion that models implied by finance theory could entirely be worthless seems rather extreme. Hence, even if the data reject the model, the decision maker may still want to use the model at least to some degree (Pastor 2000).

Some researchers involved in the mean-variance analysis of Markowitz (1952) for Portfolio Selection (PS) have only focused on PS as risk adjusted return with little or no effort being directed to the inclusion of other essential factors. Therefore, the usual portfolio analysis assumes that investors are interested only with returns attached to specific levels of risk when selecting their portfolios. In a wide variety of

R. Azmi (✉)
University of Portsmouth, Portsmouth, UK
rania.azmi@port.ac.uk, Rania.A.Azmi@gmail.com

D. Jones et al. (eds.), *New Developments in Multiple Objective and Goal Programming*, Lecture Notes in Economics and Mathematical Systems 638,
DOI 10.1007/978-3-642-10354-4_2,

applications, neither part of this restriction is desirable or important. Consequently, a portfolio analysis model that includes more essential factors in the analysis of portfolio problems is a more realistic approach. Some of these factors include liquidity, asset class, asset region, micro economics, macro economics and market dynamics.

Original PS problems, with risk and return optimisation can be viewed as a GP with two objectives. Additional objectives representing other factors can be introduced for a more realistic approach to PS problems.

Charnes et al. developed GP in 1955. GP is a multi-objective programming technique. The ethos of GP lies in the Simonan concept of satisficing of objectives (Tamiz et al. 1998). Simon introduced the concept of satisficing, a word that originated in Northumbria[1] where it meant "to satisfy". Satisficing is a strategy for making decisions in the case that one has to choose among various alternatives which are encountered sequentially, and which are not known ahead of time (Reina 2005).

GP is an important technique for decision making problems where the decision maker aims to minimize the deviation between the achievement of goals and their aspiration levels. It can be said that GP has been, and still is, the most widely used multi-objective technique in management science because of its inherent flexibility in handling decision-making problems with several conflicting objectives and incomplete or imprecise information (Romero 1991; 2004; Chang 2007).

The remaining parts of this paper are organized as follows. Section 2 discusses the literature on Multi-Criteria Decision in PS context as well as the importance of GP applications to portfolio problems. Section 3 outlines the available research papers on GP for PS. An overview of GP variants for PS is given in Sect. 4. An outline of performance measurement in portfolio selection context is provided in Sect. 5. Section 6 develops arguments for further exploitation of GP in addressing some issues in PS, and the concluding remarks are provided in Sect. 7.

2 The Use of Multi-Criteria Decision Analysis in Portfolio Selection and the Importance of Goal Programming

Optimisation is a process by which the most favourable trade-off between competing interests is determined subject to the constraints faced in any decision making process. Within the context of portfolio management, the competing interests are risk reduction and return enhancement among the other interests (Kritzman 2003).

Present-day theory of portfolio analysis prescribes a way of thinking about opportunities for investment. Instead of extensive evaluation of a single asset in isolation, the theory prescribes that investment policy can be formulated in a manner in which a purchase of an asset is done if and only if it will cause a rise in the overall

[1] A region in England on the Scottish boarder.

personal satisfactions. A rise may come about via one of three schemes as follows (Renwick 1969):

1. The new asset can cause a net increase in total present expected return on the portfolio.
2. The new asset can cause a net decline in total risk exposure on the entire portfolio.
3. There can be some subjectively acceptable trade off between change in total risk and change in total expected return on the portfolio.

The first two are the traditional and direct schemes for selecting portfolios. While, the third one is quite open to many possibilities and consequently has stimulated many studies in search for better PS for investors.

Markowitz (1952) suggests that investors should consider risk and return together and determine the allocation of funds among investment alternatives on the basis of the trade-off between them. Later on the recognition that many investors evaluate performance relative to a benchmark led to the idea of PS based on return and relative risk (Cremers et al. 2005). For many investors, both approaches fail to yield satisfactory results. Chow (1995) emphasizes that the portfolio optimisation techniques can assist in the search of portfolio that best suits each investor's particular objectives.

An alternative to Markowitz model is the Mean-Absolute Deviation (MAD) model, proposed by Konno and Yamazaki (1991). While Markowitz model assumes normality of stock returns, the MAD model does not make this assumption. The MAD model also minimizes a measure of risk, where the measure is the mean absolute deviation (Kim et al. 2005; Konno and Koshizuka 2005). Konno and Yamazaki (1991) further developed the MAD model into an equivalent GP model.

Konno and Kobayashi (1997) propose a new model for constructing an integrated stock-bond portfolio, which serves as an alternative to the popular asset allocation strategy. The fund is first allocated to indexes corresponding to diverse asset classes and then allocated to individual assets using appropriate models for each asset class.

Their model (Konno and Kobayashi 1997) determines the allocation of the fund to individual assets in one stage by solving a large scale mean-variance or mean-absolute deviation model using newly developed technologies in large scale quadratic programming and linear programming analysis, respectively. Computational experiments show that the new approach can serve as a more reliable and less expensive method to allocate the fund to diverse classes of assets.

Konno (2003) shows that there is a possibility to apply standard portfolio optimisation methods to the management of small and medium scale fund, where transaction cost and minimal transaction unit constraints are negligible. He shows that the use of mean-absolute deviation model can handle concave transaction cost and minimal transaction unit constraints in an efficient manner using branch and bound algorithm. Transaction cost is still not negligible for the majority of standard investors.

Parra et al. (2001) amongst other authors, claim that there has been a growing interest in incorporating additional criteria beyond risk and return into the PS process. Multiple criteria PS problems normally stem from multiple-argument investor utility functions. For investors with additional concerns steps can be taken to integrate them into the portfolio optimisation process more in accordance with their criteria status.

Chow (1995) mentions that investment practitioners have implicitly sent a message that optimisation models have limited relevance in real world investment decisions. One of the best arguments for this assertion is that few investors allocate their assets in the proportions indicated by an optimisation model.

Furthermore, Christiansen and Varnes (2008) present a framework for understanding how portfolio decision making is shaped through appropriate decision making. They find that the identity of decision maker is shaped and influenced by four factors, which are the formal system and rules, observations of others, the organizational context, and organizational learning. In practice, the decision maker must deal with multiple factors and criteria that make it difficult to carry out a traditional rational decision-making process.

In addition, Rifai (1996) highlights the fact that the survival of any US firm in the national and international markets depends on the use of scientific techniques in their decision-making processes. The utilization of scientific techniques requires certain steps to be followed. The most important steps are identifying, quantifying and solving the problem. He described GP as a very powerful quantitative model, which, if used properly, can be an excellent tool, particularly for investment decisions.

Cremers et al. (2005) emphasize the importance of using more approaches to portfolio formulation, particularly the mean-variance optimisation, and the full-scale optimisation approaches. They argue that institutional investors typically use mean-variance optimisation in PS, in part because it requires knowledge of only the expected returns, standard deviations, and correlations of portfolio's components, while other investors prefer to use full-scale optimisation as an alternative to mean-variance optimisation since computational advances now allow us to perform such full-scale optimisations. Under this approach, PS process considers as many asset mixes as necessary in order to identify the weights that yield the highest expected utility, given any utility function.

Renwick (1969) mentions that investment portfolio behaviour can be characterized and classified using combinations of the four interrelated variables, which are rate of return on total assets, rate of growth of output, capital structure and rate of retention of available income.

The evidence which Renwick presented in his paper (1969) supports the view that dividend policy is relevant to the investment decision as well as that finance does matter for the valuation of corporate assets. Both current and anticipated future returns on investment, along with the various types of risks associated with those returns, all interact to determine and characterize the empirical behaviour and performance of investor portfolios.

Despite the volume of research supporting standard PS, there has always been a slight undercurrent of multiple objectives in PS, but this is changing.

Generally, in PS problems the decision maker considers simultaneously conflicting objectives such as rate of return, risk and liquidity. Multi-objective programming techniques, such as GP, are used to choose the portfolio best satisfying the decision maker's aspirations and preferences.

The following figure illustrates the number of publication of research papers in the area of PS using GP:

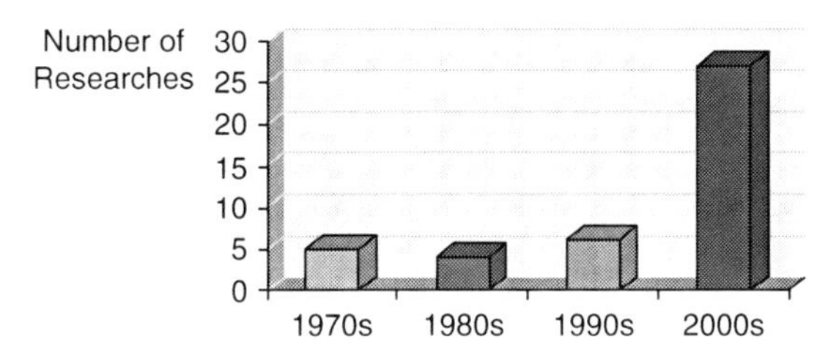

Significant advances have taken place in recent years in the field of GP. A higher level of computerized automation of the solution and modelling process has brought use of the already existing and the new analysis techniques within reach of the average practitioner (Tamiz and Jones 1998).

3 Portfolio Selection Using Goal Programming: Theoretical and Practical Developments

The ultimate objective of optimal PS is to determine and hold a portfolio which offers the minimum possible deviation for a given or desired expected return. But this objective is assuming a stable financial environment. In a world in which an investor is certain of the future, the optimal PS problem is reduced to that of structuring a portfolio that will maximize the investor's return.

Unfortunately, the future is not certain, particularly now like never before and consequently the solution to the optimal PS problem will depend upon the following elements (Callin 2008):

(a) A set of possible future scenarios for the world.
(b) A correspondence function, linking possible future scenarios to the returns of individual securities.
(c) A probabilities function of the likelihood of each of the possible future scenarios of the world.
(d) A way to determine whether one portfolio is preferable to another portfolio.

These elements are considered under different assumptions based on investors' strategy and their analysis is achievable through GP approach.

Kumar et al. (1978) highlight the fact that standard PS techniques are typically characterized by motivational assumptions of unified goals or objectives. Therefore, their immediate relevance to real-world situations, usually marked by the presence of several conflicting goals, is at best limited.

Nevertheless, with appropriate extensions the standard techniques can form the basis for accommodating multiple goals. Kumar et al. (1978) address the problem of goal conflicts in the PS of Dual-Purpose Funds, and suggest an extension of standard methodology, in terms of the development of a GP model in conceptual form, which can be applied for the resolution of inherent clash of interests.

GP in PS context is an analytical approach devised to address financial decision making problems where targets have been assigned to the attributes of a portfolio and where the decision maker is interested in minimising the non-achievement of the corresponding goals.

During the recent years many models concerning PS using GP have been developed. Amongst the research papers that introduce such models are:

Category	Year	Author	Portfolio selection using goal programming (GP variants and applications)
1970s	1973	Lee and Lerro	LGP (PS for mutual funds)
	1975	Stone and Reback	Other variant (nonlinear GP-portfolio revisions)
	1977	Booth and Dash	Other variant (nonlinear GP-bank portfolio)
	1978	Kumar et al.	LGP (dual-purpose funds)
		Muhlemann et al.	LGP (portfolio modelling)
	1979	Kumar and Philippatos	LGP (dual-purpose funds)
1980s	1980	Lee and Chesser	LGP (PS)
	1984	Levary and Avery	LGP (weighting equities in a portfolio)
	1985	Alexander and Resnich	LGP (bond portfolios)
1990s	1992	Konno and Yamazaki	MAD, PS
	1994	Byrne and Lee	Spreadsheet optimizer (real estate portfolio)
	1996	Tamiz et al.	Two staged GP model for portfolio selection
	1997	Tamiz et al.	Comparison between GP and regression analysis for PS
		Watada	Fuzzy portfolio selection
	1998	Kooros and McManis	Multiattribute optimisation for strategic investment decisions
2000s	2000	Deng et al.	Criteria, models and strategies in portfolio selection
		Inuiguchi and Ramik	Fuzzy GP and other variant
		Ogryczak	Linear programming model for portfolio selection

2001	Jobst et al.	Other variant (alternative portfolio selection models)
	Parra et al.	FGP (portfolio selection)
2002	Wang and Zhu	Fuzzy portfolio selection
	Zopounidis and Doumpos	Multi-criteria decision aid in financial decision making
2003	Allen et al.	Other variant-portfolio optimisation
	Prakash et al.	Other variant-polynomial GP (selecting portfolio with skewness)
	Rostamy et al.	Other variant
	Sun and Yan	Other variant-skewness & optimal portfolio selection
2004	Kosmidou and Zopounidis	GP, simulation analysis & bank asset liability management
	Pendaraki et al.	GP (equity mutual fund portfolios)
2005	Dash and Kajiji	Other variant (nonlinear GP for asset-liability management)
	Davies et al.	Other variant (polynomial GP-fund of hedge funds PS)
	Deng et al.	Minimax portfolio selection
	Pendaraki et al.	GP-construction of mutual funds portfolios
	Tektas et al.	GP-asset and liability management
2006	Bilbao et al.	GP-portfolio selection with expert betas
	Sharma and Sharma	LGP for mutual funds
2007	Abdelaziz et al.	Other variant
	Bilbao et al.	GP
	Gladish et al.	Interactive three-stage model-mutual funds portfolio selection
	Li and Xu	Other variant (nonlinear GP-portfolio selection)
	Sharma et al.	Other variant (credit union portfolio management)
	Wu et al.	GP-index investing

Romero claims in a recent paper (2004) that most of GP applications, which are reported in the literature, use weighted or lexicographic achievement function. He explains that this election is usually made in a rather mechanistic way without theoretical justification and if the election of the achievement function is wrong, then it is very likely that the decision maker will not accept the solution.

The majority of research papers prior to 2000 develop PS models utilising weighted and/or lexicographic GP variants. This trend has since changed to the fuzzy goal programming variant. When attributes and/or goals are in an imprecise environment and they cannot be stated with precision, it is appropriate to use fuzzy GP. A primary challenge in today's financial world is to determine how to proceed in the face of uncertainty, which arises from incomplete data and from imperfect knowledge. Volatility is an important challenge too, since estimates of volatility allow us to assess the likelihood of experiencing a particular outcome.

4 Goal Programming Variants for Portfolio Selection

Romero (2004) mentions that a key element of a GP model is the achievement function, which measures the degree of minimization of the unwanted deviational variables of the model's goals. Each type of achievement function leads to a different GP variant as follows.

4.1 Weighted Goal Programming in Portfolio Selection Models

The weighted GP for PS model usually lists the unwanted deviational variables, each weighted according to their importance.

Weighted Goal Programming (WGP) attaches weights according to the relative importance of each objective as perceived by the decision maker and minimises the sum of the unwanted weighted deviations (Tamiz and Jones 1995).

For example, the objective function in WGP model for PS seeks to minimise risk and maximise return by penalising excess risk and shortfalls in return, relative to the respective targets. Therefore, lower levels of risk and higher levels of return are not penalised. Additional objectives specifying other portfolio's attributes, such as liquidity, cost of rebalancing and sectors allocation can be included in the WGP model.

4.2 Lexicographic Goal Programming in Portfolio Selection Models

The achievement function of the lexicographic GP model to PS is made up of an ordered vector whose dimension coincides with the q number of priority levels established in the model. Each component in this vector represents the unwanted deviational variables of the goals placed in the corresponding priority level.

Lexicographic achievement functions imply a non-compensatory structure of preferences. In other words, there are no finite trade-offs among goals placed in different priority levels (Romero 2004).

The priority structure for the model can be established by assigning each goal or a set of goals to a priority level, thereby ranking the goals lexicographically in order of importance to the decision maker. When achieving one goal is equally important as achieving other goals, such as the goals of risk and return, then they may be included at the same priority level, where numerical weights represent the relative importance of the goals at the same priority level (Sharma and Sharma 2006).

Therefore, LGP could deal with many priority levels in PS problem, in which goal constraints are included according to their importance of achievement in the model. For example, a PS model using LGP may include the following priority structure:

1. Maximising the portfolio's expected return, while minimising some measurement of portfolio's risk.
2. Minimising other portfolio's risks (e.g. the systematic risk as measured by Beta coefficient).
3. Minimising the portfolio's cost of rebalancing.

+ Other priority levels.

Many authors developed lexicographic GP models for PS, particularly during the 1970s and 1980s (for example, Lee and Lerro 1973; Kumar et al. 1978; Levary and Avery 1984). Other applications of LGP for PS are developed recently within the mutual funds industry (Sharma and Sharma 2006).

4.3 MINMAX (Chebyshev) Goal Programming in Portfolio Selection Models

The achievement function of a Chebyshev GP model implies the minimization of the maximum deviation from any single goal. Moreover, when some conditions hold the corresponding solution represents a balanced allocation among the achievement of the different goals (Romero 2004).

The model of MinMax, Chebyshev, GP Portfolio Selection usually seeks the minimisation of the maximum deviation from any single goal in PS. In other words, it seeks the solution that minimizes the worst unwanted deviation from any single goal.

Some authors focus historically in developing PS models using MinMax GP variant. Deng et al. (2005), amongst others, develop MinMax GP model for PS.

4.4 Fuzzy Goal Programming in Portfolio Selection Models

While the weighted, lexicographic and MinMax forms of the achievement function are the most widely used, other recently developed variants, like Fuzzy Goal

Programming, may represent the decision makers' preferences or the decision making circumstances with more soundness.

Fuzzy mathematical programming is developed for treating uncertainties in the setting of optimisation problems. The fuzzy mathematical programming can be classified into three categories with respect to the kind of uncertainties treated in the method (Inuiguchi and Ramik 2000):

1. Fuzzy mathematical programming with vagueness.
2. Fuzzy mathematical programming with ambiguity.
3. Fuzzy mathematical programming with combined vagueness and ambiguity.

Vagueness is associated with the difficulty of making sharp or precise distinctions in the world; that is, some domain of interest is vague if it cannot be delimited by sharp boundaries, while ambiguity is associated with one-to-many relations, that is, situations in which the choice between two or more alternatives is left unspecified (Inuiguchi and Ramik 2000).

In fuzzy GP Portfolio Selection model, the decision maker is required to specify an aspiration level for each objective in the model in which aspiration levels are not known precisely. In this case, an objective with an imprecise level can be treated as a fuzzy goal (Yaghoobi and Tamiz 2006).

The use of fuzzy models not only avoids unrealistic modelling but also offers a chance for reducing information costs. Fuzzy sets are used in fuzzy mathematical programming both to define the objective and constraints and also to reflect the aspiration levels given by the decision makers (Leon et al. 2002).

In this context, Watada (1997) argues that Markowitz's approach to PS has difficulty in resolving the situation where the aspiration level and utility given by the decision makers cannot be defined exactly. Therefore, he proposes a fuzzy PS to overcome such difficulty. The fuzzy PS enables obtaining a solution which realizes the best within a vague aspiration level and the goal given as a fuzzy number, which is obtained from the expertise of the decision makers.

Gladish et al. (2007) argue that PS problem is characterized by imprecision and/or vagueness inherent in the required data, in which they proposed a three stage model, in order to mitigate such problems, based on a multi-index model and considering several market scenarios described in an imprecise way by an expert.

Gladish et al. (2007) discuss how the proposed fuzzy model allowed the decision maker to select a suitable portfolio taking into account the uncertainty related to the market scenarios and the imprecision and/or vagueness associated with the model data.

On the other hand, Leon et al. (2002) focus on the infeasible instances of different models, which are suppose to select the best portfolio according to their respective objective functions. They propose an algorithm to repair infeasibility. Such infeasibility, which usually provoked by the conflict between the desired return and the diversification requirements proposed by the investor, could be avoided by using fuzzy linear programming techniques.

Parra et al. (2001) deal with the optimum portfolio for a private investor with emphasis on three criteria, which are expected return of the portfolio, the variance return of the portfolio, and the portfolio's liquidity measured as the possibility of converting an investment into cash without any significant loss in value. They formulated these three objectives as a GP problem using fuzzy terms since they cannot be defined exactly from the point of view of the investors. Parra et al. (2001) propose a method to determine portfolios with fuzzy attributes that are set equal to fuzzy target values. Their solution is based on the investor's preferences and on the GP techniques.

Allen et al. (2003) investigate the notion of fuzziness with respect to funds allocation. They found that the boundary between the preference sets of an individual investor, for funds allocation between a risk free asset and the risky market portfolio, tends to be rather fuzzy as the investors continually evaluates and shifts their positions; unless it is a passive buy-and-hold kind of portfolio.

Inuiguchi and Ramik (2000) emphasize in their paper that the real world problems are not usually so easily formulated as mathematical models or fuzzy models. Sometimes qualitative constraints and/ or objectives are almost impossible to represent in mathematical forms. In such a situation, a fuzzy solution satisfying the given mathematical requirements are very useful in a sense of weak focus in the feasible area. Inuiguchi and Ramik (2000) applies fuzzy programming in PS problems and they found that decision maker can select the final solution from the fuzzy solution considering implicit and mathematically weak requirements.

5 Performance Measurement for Portfolios

Treynor (1965), Sharpe (1966), and Jensen (1968) developed the standard indices to measure risk adjusted returns for portfolios.

Numerous studies have tested the performance of portfolios (mutual funds) compared to a certain benchmark, usually market index, based on Sharpe, Treynor and Jensen performance measures (Artikis 2002; Cresson et al. 2002; Daniel et al. 1997; Lehmann and Modest 1987; Matallin and Nieto 2002; Otten and Schweitzer 2002; Raj et al. 2003; Zheng 1999).

Bottom-line performance measurement concentrates on the question of how a portfolio did, both absolutely and relative to a benchmark.

6 Goal Programming and Portfolio Analysis: Other Issues

The traditional portfolio optimisation model by Markowitz has not been used extensively in its original form to construct a large-scale portfolio. The first reason behind this is in the nature of the required inputs for portfolio analysis, in which accuracy is needed for returns as well as the correlation of returns. The second reason is the computational difficulty associated with solving a large-scale quadratic programming problem with dense (covariance) matrix.

Several researchers have tried to alleviate these problems by using various approximation schemes to obtain equivalent linear problems (such as Steuer et al. 2007). The use of index model reduces the amount of required computation by introducing the notion of factors influencing stock prices. However, these factors are discounted because of the popularity of equilibrium models such as the Capital Asset Pricing Model (CAPM).

The CAPM states that the expected return on a security depends only on the sensitivity of its return to the market return, its market beta. However, there is evidence that market beta does not suffice to describe expected return. In addition, the CAPM fares poorly in competition with multifactor alternatives. This evidence suggests that multifactor models should be considered in any research that requires estimates of expected returns. One popular multifactor model is the Arbitrage Pricing Theory (Fama 1996).

A factor model is not an equilibrium theory, in which it represents relationships among security returns. However, when returns are generated by a factor model, equilibrium in the capital markets will result in certain relationships among the values of the coefficients of the model. The Arbitrage Pricing Theory (APT), Like Capital Asset Pricing Models, is an equilibrium theory of the relationships between security expected returns and relevant security attributes. Unlike CAPM, the APT assumes that returns are generated by an identifiable factor model. However, it does not make strong assumptions about investor preferences (Sharpe 1985).

In order to facilitate application of his own covariance approach, Markowitz first suggested, and Sharpe (1966) later developed a market model formulation in which the rates of return on various securities are related only through common relationships with some basic underlying factor (Frankfurter and Phillip 1980).

Although GP and its variants have provided a more pragmatic tool to analyse PS problems and reach good solutions in terms of the inclusion of the decision maker's factors of importance in selecting portfolios, there are still other aspects of PS problems that can benefit from the application of GP. Some of these issues are listed and explained below.

6.1 Issues Concerning Multi-Period Returns

GP can be utilised to select portfolios on not only the basis of many factors, but also based on the future multi-period returns as well as the expected utility of multi-period returns.

Modem portfolio analysis has its origin in the work of Markowitz, who specified the portfolio problem in terms of the one-period means and variances of returns. However, most portfolio problems are multi-period. The appropriateness of one-period analysis for this class of problems has been seriously questioned in recent years. As a result, several alternative decision rules and modification of the one-period analysis have been proposed.

- Elton and Gruber (1974) evaluate two proposals that have received wide attention in the economic literature. The first involves selecting portfolios on the basis of the geometric mean of future multi-period returns. The second involves selecting portfolios on the basis of the expected utility of multi-period returns. They found that, when portfolio revision is considered, portfolio decisions based on either the expected utility of multi-period returns or the geometric mean of multi-period returns are often different from and inferior to decisions based on consideration of returns sequentially over time. This is true even when the distribution of returns is expected to be identical in each future period.
- Li and Ng (2000) consider an analytical optimal solution to the mean-variance formulation in multi-period PS. They extend the Markowitz mean-variance approach to multi-period PS problems. The derived analytical expression of the efficient frontier for the multi-period PS would enhance investors' understanding of the trade-off between the expected terminal wealth and the risk. At the same time, the derived analytical optimal multi-period portfolio policy provides investors with the best strategy to follow in a dynamic investment environment.
- Samuelson (1969) formulates and solves a many-period generalization, corresponding to lifetime planning of consumption and investment decision in his paper of Lifetime PS by Dynamic Stochastic Programming.
- Renwick (1969) emphasize almost 40 years ago that there was rarely anything even approaching unanimous agreement on any particular point of theory or interpretation of empirical data with relevance to financial analysis.

GP, if properly utilised, could provide a good approach to PS and analysis in today's complicated financial markets with multi-period returns.

6.2 Issues Concerning Extended Factors

There are a number of issues which have been introduced into practical PS problems. These include restriction on the number of assets, transaction and rebalancing costs, and cash flow or liquidity requirements.

In practice, analysts use models with both common factors, which affect all securities to a greater or lesser extent, and sector factors, which affect only some securities within a portfolio. Identification and prediction of truly pervasive factors is an extremely difficult task. Hence, the goal should be focused on permanent and important sources of security and portfolio risk and return, not the transitory and unimportant phenomena that occur in any given period (Sharpe 1985).

Nonetheless, an extended Capital Asset Pricing Models imply that expected returns may be related to additional security attributes, such as liquidity and rebalancing costs. Some of these may, in turn, be related to sensitivities to major factors.

GP and its variants provide a practical way to incorporate an extended list of factors, other than risk and return, in portfolio analysis.

For example, Steuer et al. (2007) focus on investors whose purpose is to build a suitable portfolio taking additional concerns into account. Such investors would have additional stochastic and deterministic objectives that might include dividends, number of securities in a portfolio, liquidity, social responsibility, and so forth. They develop a multiple criteria PS formulation.

Despite the acceptance and wide-spread use of the Markowitz framework, and its numerous extensions, in practice there has been a considerable debate among academics and practitioners on the validity of including only two factors for Portfolio Selection problems and equally important the validity of variance as a representative measure of risk.

6.3 Issues Concerning the Measurement of Risk

The notion of risk has found practical application within the science of Risk Management and Risk Control. Risk Control deals with limiting or eliminating specific types of risk, in as much as this is possible by taking an active position in one or more types of risk. Deciding which types of risk to mitigate is the first dilemma of a financial institution and demands considerable attention, since focusing on one particular risk category may lead to a hedged portfolio for a particular source of risk but may result in exposure to other sources of risk.

An important insight of modern financial theory is that some investment risks yield an expected reward, while other risks do not. Risks that can be eliminated by diversification do not yield an expected reward, while risks that cannot be eliminated by diversification do yield an expected reward. Thus, financial markets are somewhat fussy regarding what risks are rewarded and what risks are not (Corrado and Jordan 2005).

Diversification reduces risk, but only up to a point since some risk is diversifiable and some is not as illustrated below:

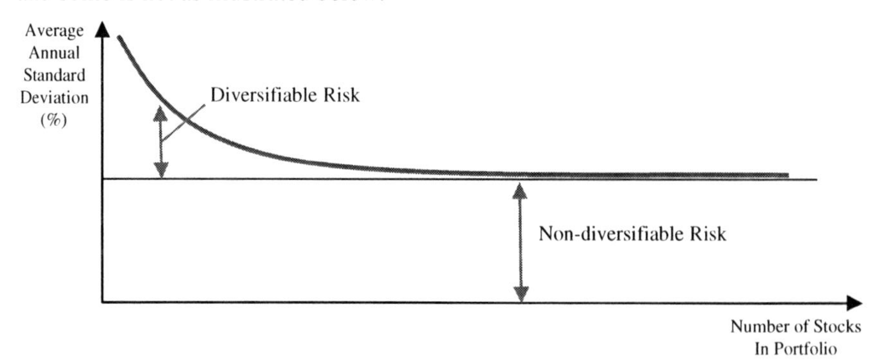

This issue becomes more challenging when optimisation models are used such as GP. For example, a GP model may result in minimisation of the risk included in the model, but the solution may be sensitive to other sources of risks that were not considered and better measured by another metric.

According to Sharpe (1966) model, the rate of return on any security is the result of two factors; a systematic component which is market related, and factors which are unique to a given security. In any application, however, concern should be not only with the alpha and beta, but with the level of uncertainty about the estimates as well.

Hu and Kercheval (2007) emphasize that portfolio optimisation requires balancing risk and return; therefore, one needs to employ some precise concept of risk. The construction of an efficient frontier depends on two inputs; a choice of risk measure, such as standard deviation, value at risk, or expected shortfall, and a probability distribution used to model returns.

Many authors provide analysis of risk measures beyond the standard deviation, such as Artzner et al. (1999), Balbas et al. (2009) and Rockafellar et al. (2006).

- For example, Mansini et al. (2007) mention that while some Linear Programs (LP) computable risk measures may be viewed as approximations to the variance (e.g., the mean absolute deviation), shortfall or quantile risk measures are recently gaining more popularity in various financial applications. Therefore, Mansini et al. (2007) study LP solvable portfolio optimization models based on extensions of the Conditional Value at Risk (CVaR) measure. The models use multiple CVaR measures thus allowing for more detailed risk aversion modeling.
- Pflug (2006) researches the measures of risk in two categories, which are risk capital measures, serve to determine the necessary amount of risk capital in order to avoid ruin if the outcomes of an economic activity are uncertain and their negative values may be interpreted as acceptability measures or safety measures, and pure risk measures, risk deviation measures, which are natural generalizations of the standard deviation.
- Rockafellar et al. (2006) study general deviation measures systematically for their potential applications to risk measurement in areas like portfolio optimization and engineering.
- Ogryczak and Ruszczynski (2002) analyse mean-risk models using quantiles and tail characteristics of the distribution. In particular, they emphasise value at risk (VAR) as a widely used quantile risk measure, which is defined as the maximum loss at a specified confidence level. Their study included also the worst conditional expectation or Tail VAR, which represents the mean shortfall at a specified confidence level.
- Artzner et al. (1999), develop a coherent measure of risk, in which they studied both market risks and nonmarket risks, and then discussed methods of measurement of these risks.

Economic analysts following the mean-variance maxim have concentrated upon the problem of portfolios as financial assets with little or no effort being directed to the inclusion of productive liabilities. Accordingly, the usual portfolio analysis assumes the absolute level of funds available for investment as fixed and concerns itself only with the distribution of that given amount over candidate opportunities. In a wide variety of applications, neither part of this restriction is either essential or desirable.

As James Tobin, the winner of the Nobel Prize in economics, showed that the investment process can be separated into two distinct steps, which are the construction of an efficient portfolio, as described by Markowitz, and the decision to combine this efficient portfolio with a riskless investment (Kritzman 2003), GP could be by far a very powerful technique for empowering the investment decision making as well as the investment process in general.

7 Conclusions

Over the last 30 years, GP for Portfolio Selection problems have been deployed extensively.

This paper has briefly reviewed many of the highlights. GP models for PS allow incorporating multiple goals such as portfolio's return, risk, liquidity, expense ratio, amongst other factors.

There is a huge capacity for future developments and applications of GP for PS issues.

In particular, GP could be used for incorporating multi-period, extended factors and different risk measures into the PS analysis. Also, the decision maker can establish target values not only for the goals but also for relevant achievement functions.

In this way, a Meta-GP model could be formulated, which allows the decision-maker to establish requirements on different achievement functions, rather than limiting their opinions to the requirements of a single variant. In this sense, this approach could be used as a second stage after GP problem for PS is being solved (Uria et al. 2002).

Future research is warrant in the area of GP applications to PS, particularly for Mutual Funds as the need for incorporating extended factors is greatly manifest.

References

Artikis P (2002) Evaluation of equity mutual funds operating in the Greek financial market. J Manag Finance 28:27–42

Artzner P, Delbaen F, Eber J-M, Heath D (1999) Coherent measures of risk. J Math Finance 9:203–228

Abdelaziz F, Aouni B, El Fayedh R (2007) Multi-objective stochastic programming for portfolio selection. Eur J Oper Res 177: 1811–1823

Alexander G, Resnick B (1985) Using linear and goal programming to immunize bond portfolios. J Banking Finance, Elsevier 9:35–54

Allen J, Bhattacharya S, Smarandache F (2003) Fuzziness and funds allocation in portfolio optimisation. Int J Soc Econ: 30(5):619–632

Balbas A, Balbas R, Mayoral S (2009) Portfolio choice and optimal hedging with general risk functions: a simplex-like algorithm. Eur J Oper Res 192:603–620

Bilbao A, Arenas M, Jimenez M, Gladish B, Rodriguez M (2006) An extension of sharpe's single-index model: portfolio selection with expert betas. J Oper Res Soc 57:1442–1451

Bilbao A, Arenas M, Rodriguez M, Antomil J (2007) On constructing expert betas for single-index model. Eur J Oper Res 183:827–847
Booth G, Dash G (1977) Bank portfolio management using non-linear goal programming. Finan Rev 14:59–69
Byrne P, Lee S (1994) Real estate portfolio analysis using a spreadsheet optimizer. J Property Finance 5(4):19–31
Callin S (2008) Portable alpha theory and practice: what investors really need to know. Wiley, Hoboken, NJ
Chang C-T (2007) Efficient structures of achievement functions for goal programming models. Asia Pac J Oper Res 24(6):755–764
Charnes A, Cooper W, Ferguson R (1955) Optimal estimation of executive compensation by linear programming. J Manag Sci 1(1):138–151
Chow G (1995) Portfolio selection based on return, risk, and relative performance. Financ Anal J March–April:54–60
Christiansen J, Varnes C (2008) From models to practice: decision making at portfolio meetings. Int J Qual Reliab Manag, 25(1):87–101
Corrado C, Jordan B (2005) Fundamentals of investments: valuations & management, 3rd edn. McGraw-Hill/Irwin, New York
Cremers J-H, Kritzman M, Page S (2005) Optimal hedge fund allocations. J Portfolio Manage Spring:70–81
Cresson J, Cudd R, Lipscomb T (2002) The early attraction of S&P 500 index funds: is perfect tracking performance an illusion? J Manag Finance 28:1–8
Daniel K, Grinblatt M, Titman S, Wermers R (1997) Measuring mutual fund performance with characteristic-based benchmarks. J Finance 52:1035–1058
Dash G, Kajiji N (2005) A nonlinear goal programming model for efficient asset-liability management of property-liability insurers. J Infor 43(2):135–156
Davies R, Kat H, Lu S (2005) Fund of hedge funds portfolio selection: a multiple objective approach. SSRN 1–31
Deng X-T, Li Z-F, Wang S-Y (2005) A minimax portfolio selection strategy with equilibrium. Eur J Oper Res 166:278–292
Deng X-T, Wang S-Y, Xia Y-S (2000) Criteria, models and strategies in portfolio selection. J Adv Model Optim (AMO) 2(2):79–103
Elton E, Gruber M (1974) On the optimality of some multiperiod portfolio selection criteria. J Bus 2:231–243
Fama E (1996) Multifactor portfolio efficiency and multifactor asset pricing. J Financ Quant Anal 31:441–465
Frankfurter G, Phillip H (1980) Portfolio selection: an analytic approach for selecting securities from a large universe. J Financ Quant Anal 15:357–377
Gladish B, Jones D, Tamiz M, Terol B (2007) An interactive three-stage model for mutual funds portfolio selection. Int J Manag Sci OMEGA 35:75–88
Hu W, Kercheval A (2007) Portfolio Optimization for Skewed-t Returns. Working paper
Inuiguchi M, Ramik J (2000) Possibilistic linear programming: a brief review of fuzzy mathematical programming and a comparison with stochastic programming in portfolio selection problem. J Fuzzy Set Syst 111:3–28
Jensen M (1968) The performance of mutual funds in the period 1945–1964. J Finance 23:389–416
Jobst N, Horniman M, Lucas C, Mitra G (2001) Computational aspects of alternative portfolio selection models in the presence of discrete asset choice constraints. J Quant Finance 1:1–13
Kim J, Kim Y, Shin K (2005) An algorithm for portfolio optimization problem. Informatica 16(1):93–106
Konno H (2003) Portfolio optimization of small scale fund using mean-absolute deviation model. Int J Theor Appl Finance 6:403–418
Konno H, Kobayashi K (1997) An integrated stock-bond portfolio optimization model. J Econ Dynam Contr 21:1427–1444
Konno H, Koshizuka T (2005) Mean-absolute deviation model. IIE Trans 37:893–900

Konno H, Yamazaki H (1991) Mean-absolute deviation portfolio optimization and its applications to Tokyo stock market. J Manag Sci 37:519–531

Kooros S, McManis B (1998) Multiattribute optimization model for strategic investment decisions. Can J Admin Sci 15(2):152–164

Kosmidou K, Zopounidis C (2004) Combining goal programming model with simulation analysis for bank asset liability management. J INFOR 42(3):175–187

Kritzman M (2003) The portable financial analyst: what practitioners need to know, 2nd edn. Wiley, New York

Kumar P, Philippatos G (1979) Conflict resolution in investment decisions: implementation of goal programming methodology for dual-purpose funds. Decis Sci 10:562–576

Kumar P, Philippatos G, Ezzell J (1978) Goal programming and selection of portfolio by dual-purpose funds. J Finance 33:303–310

Lee S, Byrne P (1998) Diversification by sector, region or function? a mean absolute deviation optimization. J Property Valuation Invest 16(1):38–56

Lee S, Chesser D (1980) Goal programming for portfolio selection. J Portfolio Manage 6:22–26

Lee S, Lerro A (1973) Optimizing the portfolio selection for mutual funds. J Finance 28:1086–1101

Lehmann B, Modest D (1987) Mutual fund performance evaluation: a comparison of benchmarks and benchmarks comparison. J Finance 42:233–265

Leon T, Liern V, Vercher E (2002) Viability of infeasible portfolio selection problems: a fuzzy approach. Eur J Oper Res 139:178–189

Levary R, Avery M (1984) On the practical application of weighting equities in a portfolio via goal programming. Opserach 21:246–261

Li D, Ng W-L (2000) Optimal dynamic portfolio selection: multiperiod mean-variance formulation. J Math Finance 10:387–406

Li J, Xu J (2007) A class of possibilistic portfolio selection model with interval coefficients and its application. J Fuzzy Optim Decis Making, Vol. 6, Springer, pp. 123–137

Mansini R, Ogryczak W, Speranza MG (2007) Conditional value at risk and related linear programming models for portfolio optimization. Ann Oper Res 152:227–256

Markowitz H (1952) Portfolio selection. J Finance 7:77–91

Matallin J, Nieto L (2002) Mutual funds as an alternative to direct stock investment. J Appl Financial Econ 743–750

Muhlemann A, Lockett A, Gear A (1978) Portfolio modelling in multiple-criteria situations under uncertainty. Decis Sci 9:612–626

Ogryczak W (2000) Multiple criteria linear programming model for portfolio selection. J Ann Oper Res, 97:143–162

Ogryczak W, Ruszczynski A (2002) Dual stochastic dominance and quantile risk measures. J Int Trans Oper Res 9:661–680

Otten R, Schweitzer M (2002) A comparison between the European and the U.S. mutual fund industry. J Manag Finance 28:14–34

Parra M, Terol A, Uria M (2001) A fuzzy goal programming approach to portfolio selection. Eur J Oper Res 133:287–297

Pastor L (2000) Portfolio selection and asset pricing models. J Finance 55:179–223

Pendaraki K, Doumpos M, Zopounidis C (2004) Towards a goal programming methodology for constructing equity mutual fund portfolios. J Asset Manag 4(6):415–428

Pendaraki K, Zopounidis C, Doumpos M (2005) On the construction of mutual fund portfolios: a multicriteria methodology and an application to the Greek market of equity mutual funds. Eur J Oper Res 163:462–481

Pflug G (2006) Subdifferential representations of risk measures. Math Program 108:339–354

Prakash A, Chang C, Pactwa T (2003) Selecting a portfolio with skewness: recent evidence from US, European, and Latin American equity markets. J Bank Finance 27:1375–1390

Raj M, Forsyth M, Tomini O (2003) Fund performance in a downside context. J Invest 12(2):50–63

Reina L (2005) From subjective expected utility theory to bounded rationality: an experimental investigation on categorization processes in integrative negotiation, in committees' decision making and in decisions under risk. Doctorate thesis, Technische Universität Dresden

Renwick F (1969) Asset management and investor portfolio behaviour: theory and practice. J Finance 24(2):180–205

Rifai A (1996) A note on the structure of the goal-programming model: assessment and evaluation. Int J Oper Prod Manag 16:40–49

Rockafellar R, Uryasev S, Zabarankin M (2006) Generalized deviations in risk analysis. Finance Stochast 10:51–74

Romero C (1991) Handbook of critical issues in goal programming. Pergamon, Oxford

Romero C (2004) A general structure of achievement function for a goal programming model. Eur J Oper Res 153:675–686

Rostamy A, Azar A, Hosseini S (2003) A mixed integer goal programming (MIGP) model for multi-period complex corporate financing problems. J Finance India 17(2):495–509

Samuelson P (1969) Lifetime portfolio selection by dynamic stochastic programming. Rev Econ Stat 51:239–246

Sharma H, Ghosh D, Sharma D (2007) Credit union portfolio management: an application of goal interval programming. Acad Bank Stud J 6(1):39–60

Sharma H, Sharma D (2006) A multi-objective decision-making approach for mutual fund portfolio. J Bus Econ Res 4:13–24

Sharpe W (1966) Mutual fund performance. J Bus 39:119–138

Sharpe W (1985) Investments, 3rd edn. Prentice-Hall, Englewood Cliffs, NJ

Steuer R, Qi Y, Hirschberger M (2007) Suitable-portfolio investors, nondominated frontier sensitivity, and the effect of multiple objectives on standard portfolio selection. J Ann Oper Res 152:297–317

Stone B, Reback R (1975) Constructing a model for managing portfolio revisions. J Bank Res 6:48–60

Sun Q, Yan Y (2003) Skewness persistence with optimal portfolio selection. J Bank Finance 27:1111–1121

Tamiz M, Hasham R, Fargher K, Jones D (1997) A comparison between goal programming and regression analysis for portfolio selection. Lect Notes Econ Math Syst 448:421–432

Tamiz M, Hasham R, Jones D (1996) A two staged goal programming model for portfolio selection. Lect Notes Econ Math Syst 432:286–299

Tamiz M, Jones D (1995) A review of goal programming and its applications. Ann Oper Res 58:39–53

Tamiz M, Jones D (1998) Goal programming: recent developments in theory and practice. Int J Manag Syst 14:1–16

Tamiz M, Jones D, Romero C (1998) Goal programming for decision making: an overview of the current state-of-the-art. Eur J Oper Res 111:569–581

Tektas A, Ozkan-Gunay E, Gunay G (2005) Asset and liability management in financial crisis. J Risk Finance 6(2):135–149

Treynor J (1965) How to rate management of investment funds. Harv Bus Rev 43:63–73

Uria MV, Caballero R, Ruiz F, Romero C (2002) Decisions aiding: meta-goal programming. Eur J Oper Res 136:422–429

Wang S, Zhu S (2002) On fuzzy portfolio selection problems. J Fuzzy Optim Decis Making 1(4):361–377

Watada J (1997) Fuzzy portfolio selection and its application to decision making. Tatra Mountains Math Publ 13:219–248

Wu L, Chou S, Yang C, Ong C (2007) Enhanced index investing based on goal programming". J Portfolio Manage 33(3):49–56

Yaghoobi M, Tamiz M (2006) On improving a weighted additive model for fuzzy goal programming problems. Int Rev Fuzzy Math 1:115–129

Zheng L (1999) Is money smart? a study of mutual fund investor's fund selection ability. J Finance 901–933

Zopounidis C, Doumpos M (2002) Multi-criteria decision aid in financial decision making: methodologies and literature review. J Multi-criteria Decis Anal 11:167–186

A Hypervolume-Based Optimizer for High-Dimensional Objective Spaces

Johannes Bader and Eckart Zitzler

Abstract In the field of evolutionary multiobjective optimization, the hypervolume indicator is the only single set quality measure that is known to be strictly monotonic with regard to Pareto dominance. This property is of high interest and relevance for multiobjective search involving a large number of objective functions. However, the high computational effort required for calculating the indicator values has so far prevented to fully exploit the potential of hypervolume-based multiobjective optimization.

This paper addresses this issue and proposes a fast search algorithm that uses Monte Carlo sampling to approximate the exact hypervolume values. In detail, we present HypE(*Hyp*ervolume *E*stimation Algorithm for Multiobjective Optimization), by which the accuracy of the estimates and the available computing resources can be traded off; thereby, not only many-objective problems become feasible with hypervolume-based search, but also the runtime can be flexibly adapted. The experimental results indicate that HypE is highly effective for many-objective problems in comparison to existing multiobjective evolutionary algorithms.

1 Motivation

By far most studies in the field of evolutionary multiobjective optimization (EMO) are concerned with the following set problem: find a set of solutions that as a whole represents a good approximation of the Pareto-optimal set. To this end, the original multiobjective problem consisting of:

- The decision space X
- The objective space $Z = \mathbb{R}^n$
- A vector function $f = (f_1, f_2, \ldots, f_n)$ comprising n objective functions $f_i : X \rightarrow \mathbb{R}$, which are without loss of generality to be minimized

J. Bader (✉)
Computer Engineering and Networks Lab, ETH Zurich, 8092 Zurich, Switzerland
e-mail: Johannes.Bader@tik.ee.ethz.ch

D. Jones et al. (eds.), *New Developments in Multiple Objective and Goal Programming*, Lecture Notes in Economics and Mathematical Systems 638,
DOI 10.1007/978-3-642-10354-4_3,

- A relation $\leqslant$ on Z, which induces a preference relation $\preceq$ on X with $a \preceq b :\Leftrightarrow f(a) \leqslant f(b)$ for $a, b \in X$

is usually transformed into a single-objective set problem (Zitzler et al. 2008).

The search space Ψ of the resulting set problem includes all possible Pareto set approximations,[1] i.e., Ψ contains all multisets over X. The preference relation $\preceq$ can be used to define a corresponding set preference relation $\preccurlyeq$ on Ψ where

$$A \preccurlyeq B :\Leftrightarrow \forall b \in B \; \exists a \in A : \; a \preceq b \tag{1}$$

for all Pareto set approximations $A, B \in \Psi$. In the following, we will assume that weak Pareto dominance is the underlying preference relation, i.e., $a \preceq b :\Leftrightarrow f(a) \leq f(b)$ (cf. Zitzler et al. 2008).[2]

A key question when tackling such a set problem is how to define the optimization criterion. Many multiobjective evolutionary algorithms (MOEAs) implement a combination of Pareto dominance on sets and a diversity measure based on Euclidean distance in the objective space, e.g., NSGA-II (Deb et al. 2000) and SPEA2 (Zitzler et al. 2002). While these methods have been successfully employed in various biobjective optimization scenarios, they appear to have difficulties when the number of objectives increases (Wagner et al. 2007). As a consequence, researchers have tried to develop alternative concepts, and a recent trend is to use set quality measures, also denoted as quality indicators, for search – so far, they have mainly been used for performance assessment. Of particular interest in this context is the hypervolume indicator (Zitzler and Thiele 1998a, 1999) as it is the only quality indicator known to be fully sensitive to Pareto dominance, i.e., whenever a set of solutions dominates another set, it has a higher hypervolume indicator value than the second set. This property is especially desirable when many objective functions are involved.

Several hypervolume-based MOEAs have been proposed meanwhile (e.g., Emmerich et al. 2005; Igel et al. 2007; Brockhoff and Zitzler 2007), but their main drawback is their extreme computational overhead. Although there have been recent studies presenting improved algorithms for hypervolume calculation, currently high-dimensional problems with six or more objectives are infeasible for these MOEAs. Therefore, the question is whether and how fast hypervolume-based search algorithms can be designed that exploit the advantages of the hypervolume indicator and at the same time are scalable with respect to the number of objectives.

[1] Here, a Pareto set approximation may also contain dominated solutions as well as duplicates, in contrast to the notation in Zitzler et al. (2003).

[2] For reasons of simplicity, we will use the term "u weakly dominates v" resp. "u dominates v" independently of whether u and v are elements of X, Z, or Ψ. For instance, A weakly dominates b with $A \in \Psi$ and $b \in X$ means $A \preccurlyeq \{b\}$ and a dominates z with $a \in X$ and $z \in Z$ means $f(a) \leq z \wedge z \not\leq f(a)$.

2 Related Work

The hypervolume indicator was originally proposed and employed in Zitzler and Thiele (1998, 1999) to compare quantitatively the outcomes of different MOEAs. In these two first publications, the indicator was denoted as "size of the space covered", and later also other terms such as "hyperarea metric" (Van Veldhuizen 1999), "S-metric" (Zitzler 1999), "hypervolume indicator" (Zitzler et al. 2003), and "hypervolume measure" (Beume et al. 2007) were used. Besides the names, there are also different definitions available, based on polytopes (Zitzler and Thiele 1999), the attainment function (Zitzler et al. 2007), or the Lebesgue measure (Laumanns et al. 1999; Knowles 2002; Fleischer 2003).

Knowles (2002) and Knowles and Corne (2003) were the first to propose the integration of the hypervolume indicator into the optimization process. In particular, they described a strategy to maintain a separate, bounded archive of nondominated solutions based on the hypervolume indicator. Huband et al. (2003) presented an MOEA which includes a modified SPEA2 environmental selection procedure where a hypervolume-related measure replaces the original density estimation technique. In Zitzler and Künzli (2004), the binary hypervolume indicator was used to compare individuals and to assign corresponding fitness values within a general indicator-based evolutionary algorithm (IBEA). The first MOEA tailored specifically to the hypervolume indicator was described in Emmerich et al. (2005); it combines nondominated sorting with the hypervolume indicator and considers one offspring per generation (steady state). Similar fitness assignment strategies were later adopted in Zitzler et al. (2007) and Igel et al. (2007), and also other search algorithms were proposed where the hypervolume indicator is partially used for search guidance (Nicolini 2005; Mostaghim et al. 2007). Moreover, specific aspects like hypervolume-based environmental selection (Bradstreet et al. 2006), cf. Sect. 3.3, and explicit gradient determination for hypervolume landscapes (Emmerich et al. 2007) have been investigated recently.

The major drawback of the hypervolume indicator is its high computation effort; all known algorithms have a worst-case runtime complexity that is exponential in the number of objectives, more specifically $\mathcal{O}(N^{n-1})$ where N is the number of solutions considered (Knowles 2002; While et al. 2006). A different approach was presented by Fleischer (2003) who mistakenly claimed a polynomial worst-case runtime complexity – While (2005) showed that it is exponential in n as well. Recently, advanced algorithms for hypervolume calculation have been proposed, a dimension-sweep method (Fonseca et al. 2006) with a worst-case runtime complexity of $\mathcal{O}(N^{n-2} \log N)$, and a specialized algorithm related to the Klee measure problem (Beume and Rudolph, 2006) the runtime of which is in the worst case of order $\mathcal{O}(N \log N + N^{n/2})$. Furthermore, Yang and Ding (2007) described an algorithm for which they claim a worst-case runtime complexity of $\mathcal{O}((n/2)^N)$. The fact that there is no exact polynomial algorithm available gave rise to the hypothesis that this problem in general is hard to solve, although the tightest known lower bound is of order $\Omega(N \log N)$ (Beume et al. 2007a). New results substantiate this hypothesis:

Bringmann and Friedrich (2008) have proven that the problem of computing the hypervolume is $\#P$-complete, i.e., it is expected that no polynomial algorithm exists since this would imply $NP = P$.

The issue of speeding up the hypervolume indicator has been addressed in different ways: by automatically reducing the number of objectives (Brockhoff and Zitzler 2007) and by approximating the indicator values using Monte Carlo simulation (Everson et al. 2002; Bader et al. 2008; Bringmann and Friedrich 2008). Everson et al. (2002) used a basic Monte Carlo technique for performance assessment in order to estimate the values of the binary hypervolume indicator (Wagner et al. 2007); with their approach the error ratio is not polynomially bounded. In contrast, the scheme presented in Bringmann and Friedrich (2008) is a fully polynomial randomized approximation scheme where the error ratio is polynomial in the input size. Another study (Bader et al. 2008) – a precursor study for the present paper – employed Monte Carlo sampling for fast hypervolume-based search. The main idea is to estimate – by means of Monte Carlo simulation – the ranking of the individuals that is induced by the hypervolume indicator and not to determine the exact indicator values. This paper proposes an advanced method called HypE (*Hyp*ervolume *E*stimation Algorithm for Multiobjective Optimization) that is based on the same idea, but uses a novel fitness assignment scheme for both mating and environmental selection, that can be effectively approximated.

As we will show in the following, the proposed search algorithm can be easily tuned regarding the available computing resources and the number of objectives involved. Thereby, it opens a new perspective on how to treat many-objective problems, and the presented concepts may also be helpful for other types of quality indicators to be integrated in the optimization process.

3 HypE: Hypervolume Estimation Algorithm for Multiobjective Optimization

When considering the hypervolume indicator as the objective function of the underlying set problem, the main question is how to make use of this measure within a multiobjective optimizer to guide the search. In the context of a MOEA, this refers to selection and one can distinguish two situations:

1. The selection of solutions to be varied (mating selection)
2. The selection of solutions to be kept in the population (environmental selection)

In the following, we outline a new algorithm based on the hypervolume indicator called HypE. Thereafter, the two selection steps mentioned above as realized in HypE are presented in detail.

3.1 Algorithm

HypE belongs to the class of simple indicator-based evolutionary algorithm, as for instance discussed in Zitzler et al. (2007). As outlined in Algorithm 1, HypE reflects

Algorithm 1 HypE Main Loop

Require: reference set $R \subseteq Z$, population size $N \in \mathbb{N}$, number of generations $g_{\max}$, number of sampling points $M \in \mathbb{N}$

1: initialize population P by selecting N solutions from X uniformly at random
2: $g \leftarrow 0$
3: **while** $g \leq g_{\max}$ **do**
4: $\quad P' \leftarrow \mathit{matingSelection}(P, R, N, M)$
5: $\quad P'' \leftarrow \mathit{variation}(P', N)$
6: $\quad P \leftarrow \mathit{environmentalSelection}(P \cup P'', R, N, M)$
7: $\quad g \leftarrow g + 1$

a standard evolutionary algorithm that consists of the successive application of mating selection, variation, and environmental selection. As to mating selection, binary tournament selection is proposed here, although any other selection scheme could be used as well, where the tournament selection is based on the fitness proposed in Sect. 3.2. The procedure *variation* encapsulates the application of mutation and recombination operators to generate N offspring. Finally, environmental selection aims at selecting the most promising N solutions from the multiset-union of parent population and offspring; more precisely, it creates a new population by carrying out the following two steps:

1. First, the union of parents and offspring is divided into disjoint partitions using the principle of nondominated sorting (Goldberg 1989; Deb et al. 2000), also known as dominance depth. Starting with the lowest dominance depth level, the partitions are moved one by one to the new population as long as the first partition is reached that cannot be transferred completely. This corresponds to the scheme used in most hypervolume-based multiobjective optimizers (Emmerich et al. 2005; Igel et al. 2007; Brockhoff and Zitzler 2007).
2. The partition that only fits partially into the new population is then processed using the novel fitness scheme presented in Sect. 3.3. In each step, the fitness values for the partition under consideration are computed and the individual with the worst fitness is removed – if multiple individuals share the same minimal fitness, then one of them is selected uniformly at random. This procedure is repeated until the partition has been reduced to the desired size, i.e., until it fits into the remaining slots left in the new population.

The scheme of first applying non-dominated sorting is similar to other algorithms (e.g., Igel et al. 2007; Emmerich et al. 2005). The differences are: (1) the fitness assignment scheme for mating, (2) the one for environmental selection, and (3) the method how the fitness values are determined. The estimation of the fitness values by means of Monte Carlo sampling is discussed in Sect. 3.4.

3.2 Basic Scheme for Mating Selection

To begin with, we formally define the hypervolume indicator as a basis for the following discussions. Different definitions can be found in the literature, and we here use the one from Zitzler et al. (2008) which draws upon the Lebesgue measure as proposed in Laumanns et al. (1999) and considers a reference set of objective vectors.

Definition 1. Let $A \in \Psi$ be a Pareto set approximation and $R \subset Z$ be a reference set of mutually nondominating objective vectors. Then the hypervolume indicator I_H can be defined as

$$I_H(A, R) := \lambda(H(A, R)), \tag{2}$$

where

$$H(A, R) := \{z \in Z \,;\, \exists a \in A \, \exists r \in R : f(a) \leq z \leq r\} \tag{3}$$

and λ is the Lebesgue measure with $\lambda(H(A, R)) = \int_{\mathbb{R}^n} \mathbf{1}_{H(A,R)}(z)dz$ and $\mathbf{1}_{H(A,R)}$ being the characteristic function of $H(A, R)$.

The set $H(A, R)$ denotes the set of objective vectors that are enclosed by the front $f(A)$ given by A and the reference set R. It can be further split into partitions $H(S, A, R)$, each associated with a specific subset $S \subseteq A$:

$$H(S, A, R) := \Big[\bigcap_{s \in S} H(\{s\}, R)\Big] \setminus \Big[\bigcup_{a \in A \setminus S} H(\{a\}, R)\Big]. \tag{4}$$

The set $H(S, A, R) \subseteq Z$ represents the portion of the objective space that is jointly weakly dominated by the solutions in S and not weakly dominated by any other solution in A. The partitions $H(S, A, R)$ are disjoint and the union of all partitions is $H(A, R)$ which is illustrated in Fig. 1.

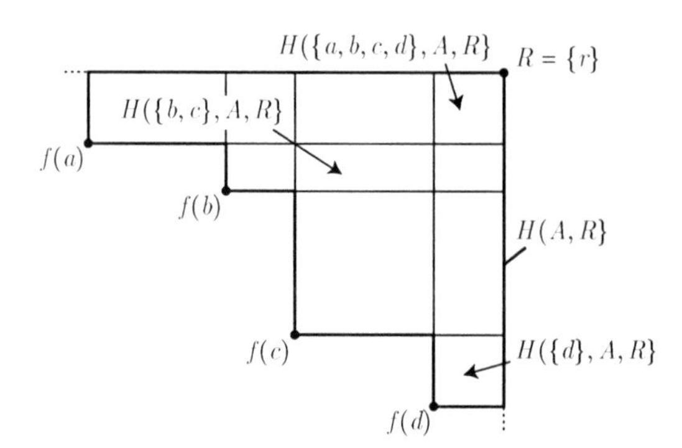

Fig. 1 Illustration of the notions of $H(A, R)$ and $H(S, A, R)$ in the objective space for a Pareto set approximation $A = \{a, b, c, d\}$ and reference set $R = \{r\}$

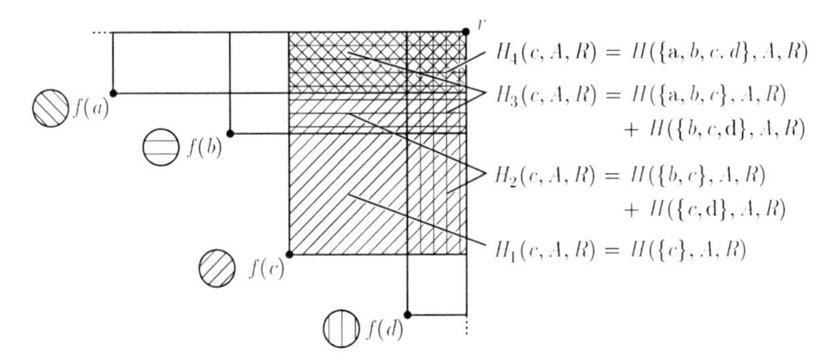

Fig. 2 Illustration of the notions of $H(A, R)$ and $H_i(a, A, R)$ in the objective space for a Pareto set approximation $A = \{a, b, c, d\}$ and reference set $R = \{r\}$

In practice, it is infeasible to determine all distinct $H(S, A, R)$ due to combinatorial explosion. Instead, we will consider a more compact splitting of the dominated objective space that refers to single solutions:

$$H_i(a, A, R) := \bigcup_{\substack{S \subseteq A \\ a \in S \\ |S| = i}} H(S, A, R). \tag{5}$$

According to this definition, $H_i(a, A, R)$ stands for the portion of the objective space that is jointly and solely weakly dominated by a and any $i-1$ further solutions from A, see Fig. 2. Note that the sets $H_1(a, A, R), H_2(a, A, R), \ldots, H_{|A|}(a, A, R)$ are disjoint for a given $a \in A$ while the sets $H_i(a, A, R)$ and $H_i(b, A, R)$ may be overlapping for fixed i and different solutions $a, b \in A$. This slightly different notion has reduced the number of subspaces to be considered from $2^{|A|}$ for $H(S, A, R)$ to $|A|^2$ for $H_i(a, A, R)$.

Now, given an arbitrary population $P \in \Psi$ one obtains for each solution a contained in P a vector $(\lambda(H_1(a, P, R)), \lambda(H_2(a, P, R)), \ldots, \lambda(H_{|P|}(a, P, R)))$ of hypervolume contributions. These vectors can be used to assign fitness values to solutions; while most hypervolume-based search algorithms only take the first components, i.e., $\lambda(H_1(a, P, R))$, into account, we here propose the following scheme to aggregate the hypervolume contributions into a single scalar value.

Definition 2. Let $A \in \Psi$ and $R \subset Z$. Then the function I_h with

$$I_h(a, A, R) := \sum_{i=1}^{|A|} \frac{1}{i} \lambda(H_i(a, A, R)) \tag{6}$$

gives for each solution $a \in A$ the hypervolume that can be attributed to a with regard to the overall hypervolume $I_H(A, R)$.

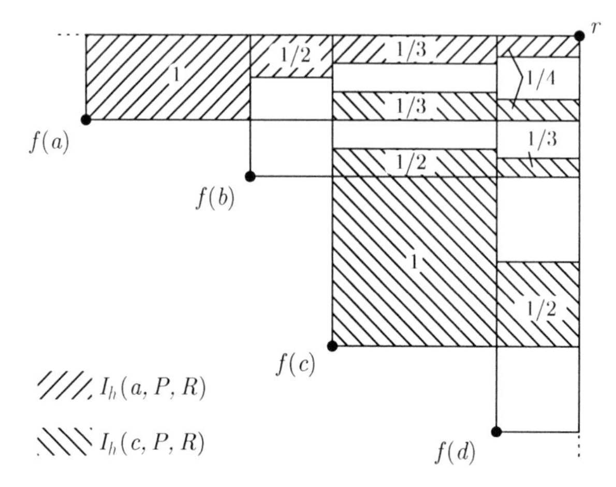

Fig. 3 Illustration of the basic fitness assignment scheme. The fitness of two individuals a and c of a Pareto set approximation $A = \{a, b, c, d\}$ is set to $F_a = I_h(a, A, R)$ and $F_b = I_h(b, A, R)$ respectively (*diagonally hatched areas*)

The motivation behind this definition is simple: the hypervolume contribution of each partition $H(S, A, R)$ is shared equally among the dominating solutions $s \in S$. That means the portion of Z solely weakly dominated by a specific solution a is fully attributed to a, the portion of Z that a weakly dominates together with another solution b is attributed half to a and so forth – the principle is illustrated in Fig. 3. Thereby, the overall hypervolume is distributed among the distinct solutions according to their hypervolume contributions as the following theorem shows (the proof can be found in the appendix). Note that this scheme does not require that the solutions of the considered Pareto set approximation A are mutually non-dominating; it applies to nondominated and dominated solutions alike.

Next, we will extend and generalize the fitness assignment scheme with regard to the environmental selection phase.

3.3 Extended Scheme for Environmental Selection

In EMO, environmental selection is mostly carried out by first merging parents and offspring and then truncating the resulting union by choosing the subset that represents the best Pareto set approximation. The number k of solutions to be achieved usually equals the population size, and therefore the exact computation of the best subset is computationally infeasible. Instead, the optimal subset is approximated in terms of a greedy heuristic (Zitzler and Künzli 2004; Brockhoff and Zitzler 2007): all solutions are evaluated with respect to their usefulness and the least important solution is removed; this process is repeated until k solutions have been removed.

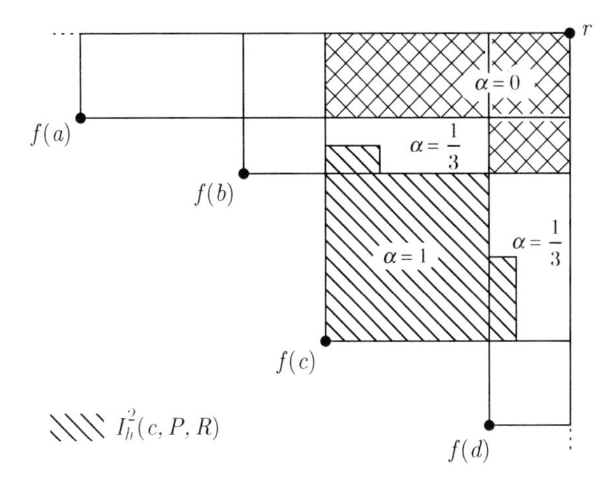

Fig. 4 The figure shows for $A = \{a, b, c, d\}$ and $R = \{r\}$ (1) which portion of the objective space remains dominated if any two solutions are removed from A (*shaded area*), and (2) the probabilities α that a particular area that can be attributed to $a \in A$ is lost if a is removed from A together with any other solution in A

The key issue with respect to the above greedy strategy is how to evaluate the usefulness of a solution. The scheme presented in Definition 2 has the drawback that portions of the objective space are taken into account that for sure will not change. Suppose, for instance, a population with four solutions as shown in Fig. 4; when two solutions need to be removed ($k = 2$), then the subspaces $H(\{a, b, c\}, P, R)$, $H(\{b, c, d\}, P, R)$, and $H(\{a, b, c, d\}, P, R)$ remain weakly dominated independently of which solutions are deleted. This observation led to the idea of considering the expected loss in hypervolume that can be attributed to a particular solution when exactly k solutions are removed. In detail, we consider for each $a \in P$ the average hypervolume loss over all subsets $S \subseteq P$ that contain a and $k - 1$ further solutions; this value can be easily computed by slightly extending the scheme from Definition 2 as follows.

Definition 3. Let $A \in \Psi$, $R \subset Z$, and $k \in \{0, 1, \ldots, |A|\}$. Then the function I_h^k with

$$I_h^k(a, A, R) := \sum_{i=1}^{k} \frac{\alpha_i}{i} \lambda(H_i(a, A, R)), \tag{7}$$

where

$$\alpha_i := \prod_{j=1}^{i-1} \frac{k - j}{|A| - j} \tag{8}$$

gives for each solution $a \in A$ the expected hypervolume loss that can be attributed to a when a and $k - 1$ uniformly randomly chosen solutions from A are removed from A.

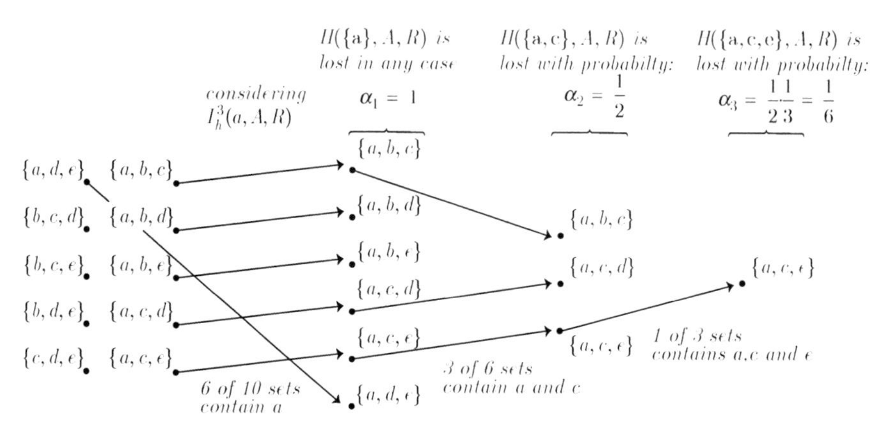

Fig. 5 Illustrates the calculation of α_i when removing a and two other individuals from the set $A = \{a, b, c, d, e\}$. Each subset of three individuals that contains a is considered equally likely; therefore, the chance of losing a and c is 3/6 and the probability of losing a, c, and e is (1/2)(1/3)

The correctness of (7) can be proved by mathematical induction (Bader and Zitzler 2008). In the following example we illustrates the calculation of α_i to put the idea of expected loss across:

Example 1. Out of five individuals $A = \{a, b, c, d, e\}$, three have to be removed and we want to know $I_h^3(a, A, R)$ (see Fig. 5). To this end, the expected shares of $H_i(a, A, R)$ that are lost when removing a – represented by α_i – have to be determined. The first coefficient is $\alpha_1 = 1$, because $H_1(a, A, R)$ is lost for sure, since the space is only dominated by a. Calculating the probability of losing the partitions dominated by a and a second individual $H_2(a, A, R)$ is a little bit more involved. Without loss of generality we consider the space dominated only by a and c. In addition to a, there are two individuals to be removed out of the remaining four individuals. Since we assume, by approximation every removal is equally probable, the chance of removing c and thereby losing the partition $H(\{a, c\}, A, R)$ is $\alpha_2 = 1/2$. Finally, for α_3 three points including a have to be taken off. The probability of removing a and any other individual was just derived to be 1/2. Out of three remaining individuals, the chance to remove one particular, say e, is 1/3. This gives the coefficient $\alpha_3 = (1/2)(1/3)$. For reasons of symmetry, these calculations hold for any other partition dominated by a and $i - 1$ points. Therefore, the expected loss of $H_i(a, A, R)$ is $\alpha_i \lambda(H_i(a, A, R))$. Since i points share the partition $H_i(a, A, R)$, the expected loss is additionally multiplied by $1/i$.

Notice that $I_h^1(a, A, R) = \lambda(H_1(a, A, R))$ and $I_h^{|A|}(a, A, R) = I_h(a, A, R)$, i.e., this modified scheme can be regarded as a generalization of the scheme presented in Definition 2 and the commonly used fitness assignment strategy for hypervolume-based search (Knowles and Corne 2003; Emmerich et al. 2005; Igel et al. 2007; Bader et al. 2008).

The fitness values $I_h^k(a, A, R)$ can be calculated on the fly for all individuals $a \in A$ by a slightly extended version of the "hypervolume by slicing objectives" (Zitzler 2001; Knowles 2002; While et al. 2006) algorithm, which traverse recursively one objective after another. It differs from existing methods in that it allows (1) to consider a set R of reference points and (5) to compute all fitness values, e.g., the $I_h^1(a, P, R)$ values for $k = 1$, in parallel for any number of objectives instead of subsequently as in Beume et al. (2007). Basically, the modification concerns also considering solutions, which are dominated according to a particular scanline. A detailed description of the extended algorithm can be found in Bader and Zitzler (2008).

Unfortunately, the worst case complexity of the algorithm is $\mathcal{O}(|A|^n)$ which renders the exact calculation inapplicable for problems with more than about five objectives. However, in the context of randomized search heuristics one may argue that the exact fitness values are not crucial and approximated values may be sufficient. These considerations lead to the idea of estimating the fitness values by Monte Carlo sampling, whose basic principle is described in the following section.

3.4 *Estimating the Fitness Values Using Monte Carlo Sampling*

To approximate the fitness values according to Definitions 2 and 3, we need to estimate the Lebesgue measures of the domains $H_i(a, P, R)$ where $P \in \Psi$ is the population. Since these domains are all integrable, their Lebesgue measure can be approximated by means of Monte Carlo simulation.

For this purpose, a sampling space $S \subseteq Z$ has to be defined with the following properties: (1) the hypervolume of S can easily be computed, (2) samples from the space S can be generated fast, and (3) S is a superset of the domains $H_i(a, P, R)$ the hypervolumes of which one would like to approximate. The latter condition is met by setting $S = H(P, R)$, but since it is hard both to calculate the Lebesgue measure of this sampling space and to draw samples from it, we propose using the axis-aligned minimum bounding box containing the $H_i(a, P, R)$ subspaces instead, i.e.:

$$S := \{(z_1, \ldots, z_n) \in Z \mid \forall 1 \leq i \leq n : l_i \leq z_i \leq u_i\}, \tag{9}$$

where

$$\begin{aligned} l_i &:= \min_{a \in P} f_i(a) \\ u_i &:= \max_{(r_1,\ldots,r_n) \in R} r_i \end{aligned} \tag{10}$$

for $1 \leq i \leq n$. Hence, the volume V of the sampling space S is given by $V = \prod_{i=1}^n \max\{0, u_i - l_i\}$.

Now given S, sampling is carried out by selecting M objective vectors $s_1, \ldots, s_M$ from S uniformly at random. For each s_j it is checked whether it lies in any partition $H_i(a, P, R)$ for $1 \leq i \leq k$ and $a \in P$. This can be determined in two steps: first, it is verified that s_j is "below" the reference set R, i.e., it exists $r \in R$ that is dominated by s_j; second, it is verified that the multiset A of those population members dominating s_j is not empty. If both conditions are fulfilled, then we know that – given A – the sampling point s_j lies in all partitions $H_i(a, P, R)$ where $i = |A|$ and $a \in A$. This situation will be denoted as a *hit* regarding the ith partition of a. If any of the above two conditions is not fulfilled, then we call s_j a *miss*. Let $X_j^{(i,a)}$ denote the corresponding random variable that is equal to 1 in case of a hit of s_j regarding the ith partition of a and 0 otherwise.

Based on the M sampling points, we obtain an estimate for $\lambda(H_i(a, P, R))$ by simply counting the number of hits and multiplying the hit ratio with the volume of the sampling box:

$$\hat{\lambda}\big(H_i(a, P, R)\big) = \frac{\sum_{j=1}^{M} X_j^{(i,a))}}{M} V. \tag{11}$$

This value approaches the exact value $\lambda(H_i(a, P, R))$ with increasing M by the law of large numbers. Due to the linearity of the expectation operator, the fitness scheme according to (7) can be approximated by replacing the Lebesgue measure with the respective estimates given by (11):

$$\hat{I}_h^k(a, P, R) = \sum_{i=1}^{k} \frac{\alpha_i}{i} \left(\frac{\sum_{j=1}^{M} X_j^{(i,a))}}{M} V \right). \tag{12}$$

Note that the partitions $H_i(a, P, R)$ with $i > k$ do not need to be considered for the fitness calculation as they do not contribute to the I_h^k values that we would like to estimate, cf. Definition 3.

4 Experiments

4.1 Experimental Setup

HypE is implemented within the PISA framework (Bleuler et al. 2003) and tested in two versions: the first, denoted by HypE, uses fitness-based mating selection as described in Sect. 3.2, while the second, HypE*, employs a uniform mating selection scheme where all individuals have the same probability of being chosen for reproduction. Unless stated otherwise, for sampling the number of sampling points is fixed to $M = 10{,}000$ and kept constant during a run.

On the one hand, HypE and HypE* are compared to three popular MOEAs, namely NSGA-II (Deb et al. 2000), SPEA2 (Zitzler et al. 2002), and IBEA (in combination with the ε-indicator) (Zitzler and Künzli 2004). Since these algorithms are not designed to optimize the hypervolume, it cannot be expected that they perform particularly well when measuring the quality of the approximation in terms of the hypervolume indicator. Nevertheless, they serve as an important reference as they are considerably faster than hypervolume-based search algorithms and therefore can execute a substantially larger number of generations when keeping the available computation time fixed. On the other hand, we include the sampling-based optimizer proposed in Bader et al. (2008), here denoted as SHV (sampling-based hypervolume-oriented algorithm); finally, to study the influence of the nondominated sorting we also include a simple HypE variant named RS (random selection) where all individuals of the same dominance depth level are assigned the same constant fitness value. Thereby, the selection pressure is only maintained by the nondominated sorting carried out during the environmental selection phase.

In this paper, we focus on many-objective problems for which exact hypervolume-based methods (e.g., Emmerich et al. 2005; Igel et al. 2007) are not applicable. Therefore, we did not include these algorithms in the experiments. However, the interested reader is referred to Bader and Zitzler (2008) where HypE is compared to an exact hypervolume algorithm.

As basis for the comparisons, the DTLZ (Deb et al. 2005), and the WFG (Huband et al. 2006) test problem suites are considered since they allow the number of objectives to be scaled arbitrarily – here, ranging from 2 to 50 objectives. For the DTLZ problem, the number of decision variables is set to 300, while for the WFG problems individual values are used.[3]

The individuals are represented by real vectors, where a polynomial distribution is used for mutation and the SBX-20 operator is used for recombination (Deb 2001). The recombination and mutation probabilities are set according to Deb et al. (2005).

For each benchmark function, 30 runs are carried out per algorithm using a population size of $N = 50$. Either the maximum number of generations was set to $g_{\max} = 200$ (results shown in Table 1) or the runtime was fixed to 30 min for each run (results shown in Fig. 6). For each run, the hypervolume of the last population is determined, where for less than six objectives these are calculated exactly and otherwise approximated by Monte Carlo sampling. For each algorithm A_i, the hypervolume values are then subsumed under the performance score $P(A_i)$, which represents the number of other algorithms that achieved significantly higher hypervolume values on the particular test case. The test for significance is done using Kruskal–Wallis and the Conover Inman post hoc tests (Conover 1999). For a full description of the performance score, please see Zamora and Burguete (2008).

4.2 Results

Table 1 shows the performance score and mean hypervolume of the different algorithms on six test problems. In 18 instances, HypE is better than HypE*, while vice versa HypE* is better than HypE only in four cases. HypE reaches the best performance score overall. Summing up all performance scores, HypE yields the best

[3] The number of decision variables (first value in parenthesis) and their decomposition into position (second value) and distance variables (third value) as used by the WFG test function for different number of objectives are: 2d (24, 4, 20); 3d (24, 4, 20); 5d (50, 8, 42); 7d (70, 12, 58); 10d (59,9,50); 25d (100, 24, 76); 50d (199,49,150).

Table 1 Comparison of HypE to different EAs for the hypervolume indicator, where n denotes the number of objectives. The first number represents the performance score $P(A_i)$, which stands for the number of contenders significantly dominating algorithm A_i. The number in brackets denote the hypervolume value (to be maximized), normalized to the minimum and maximum value observed on the testproblem

	n	SHV	IBEA	NSGA-II	RS	SPEA2	HypE	HypE*
	2	2 (0.438)	0 (0.871)	5 (0.306)	5 (0.278)	2 (0.431)	1 (0.682)	4 (0.362)
	3	2 (0.995)	0 (0.998)	5 (0.683)	6 (0.491)	4 (0.888)	1 (0.996)	3 (0.994)
	5	1 (0.998)	0 (0.999)	4 (0.808)	6 (0.324)	5 (0.795)	2 (0.998)	3 (0.998)
DTLZ 2	7	3 (0.998)	0 (1.000)	5 (0.808)	6 (0.340)	4 (0.850)	1 (0.999)	1 (0.999)
	10	3 (0.999)	2 (1.000)	5 (0.825)	6 (0.290)	4 (0.868)	0 (1.000)	0 (1.000)
	25	3 (0.999)	2 (1.000)	4 (0.965)	6 (0.301)	5 (0.882)	0 (1.000)	0 (1.000)
	50	3 (1.000)	2 (1.000)	4 (0.998)	6 (0.375)	5 (0.917)	0 (1.000)	0 (1.000)
	2	1 (0.848)	0 (0.928)	3 (0.732)	3 (0.834)	2 (0.769)	1 (0.779)	3 (0.711)
	3	1 (0.945)	0 (0.989)	3 (0.777)	3 (0.774)	2 (0.860)	0 (0.987)	2 (0.922)
	5	1 (0.997)	0 (0.998)	4 (0.749)	5 (0.558)	6 (0.537)	2 (0.992)	2 (0.992)
DTLZ 4	7	1 (0.999)	0 (1.000)	4 (0.902)	6 (0.569)	5 (0.814)	2 (0.999)	2 (0.999)
	10	2 (1.000)	0 (1.000)	4 (0.988)	6 (0.560)	5 (0.960)	1 (1.000)	0 (1.000)
	25	3 (1.000)	2 (1.000)	4 (1.000)	6 (0.546)	5 (0.991)	0 (1.000)	0 (1.000)
	50	2 (1.000)	2 (1.000)	4 (1.000)	6 (0.517)	5 (0.999)	0 (1.000)	0 (1.000)
	2	0 (0.945)	1 (0.898)	6 (0.739)	2 (0.818)	4 (0.817)	2 (0.853)	1 (0.876)
	3	0 (0.993)	1 (0.987)	6 (0.633)	4 (0.794)	5 (0.722)	3 (0.970)	2 (0.980)
	5	0 (0.988)	0 (0.986)	6 (0.478)	4 (0.672)	5 (0.569)	2 (0.868)	2 (0.862)
DTLZ 7	7	0 (0.981)	1 (0.958)	5 (0.348)	4 (0.559)	5 (0.352)	2 (0.877)	2 (0.870)
	10	0 (0.986)	1 (0.831)	4 (0.137)	6 (0.057)	4 (0.166)	2 (0.744)	1 (0.781)
	25	0 (0.973)	0 (0.966)	3 (0.856)	2 (0.893)	4 (0.671)	2 (0.889)	3 (0.825)
	50	1 (0.767)	0 (0.966)	5 (0.233)	4 (0.254)	6 (0.020)	2 (0.684)	3 (0.675)

Table 1 (continued)

	2	4 (0.567)	0 (0.949)	1 (0.792)	6 (0.160)	1 (0.776)	2 (0.744)	4 (0.557)
	3	4 (0.792)	3 (0.811)	3 (0.827)	6 (0.207)	1 (0.881)	0 (0.985)	1 (0.894)
	5	4 (0.766)	5 (0.703)	2 (0.832)	6 (0.291)	2 (0.820)	0 (0.973)	1 (0.951)
WFG 1	7	4 (0.647)	5 (0.649)	2 (0.814)	6 (0.189)	2 (0.812)	0 (0.956)	1 (0.937)
	10	6 (0.402)	4 (0.843)	2 (0.932)	5 (0.562)	2 (0.937)	0 (0.977)	0 (0.975)
	25	6 (0.183)	4 (0.930)	0 (0.971)	5 (0.815)	3 (0.965)	0 (0.972)	0 (0.973)
	50	6 (0.210)	4 (0.869)	2 (0.962)	4 (0.823)	2 (0.961)	0 (0.971)	0 (0.970)
	2	1 (0.987)	4 (0.962)	3 (0.974)	6 (0.702)	4 (0.969)	0 (0.990)	0 (0.989)
	3	0 (0.556)	3 (0.475)	3 (0.406)	6 (0.261)	2 (0.441)	0 (0.446)	0 (0.372)
	5	0 (0.671)	0 (0.533)	0 (0.644)	6 (0.351)	0 (0.624)	0 (0.557)	3 (0.503)
WFG 2	7	0 (0.632)	0 (0.747)	1 (0.409)	5 (0.155)	0 (0.837)	0 (0.528)	0 (0.630)
	10	0 (0.971)	0 (0.988)	0 (0.978)	5 (0.020)	2 (0.962)	0 (0.981)	1 (0.966)
	25	0 (0.951)	0 (0.951)	2 (0.935)	6 (0.072)	2 (0.933)	2 (0.934)	2 (0.928)
	50	3 (0.538)	0 (0.962)	0 (0.959)	6 (0.076)	0 (0.952)	2 (0.945)	3 (0.943)
	2	2 (0.994)	0 (0.997)	4 (0.991)	6 (0.559)	4 (0.990)	0 (0.997)	2 (0.994)
	3	2 (0.995)	3 (0.981)	4 (0.966)	6 (0.689)	4 (0.966)	0 (0.999)	1 (0.998)
	5	6 (0.339)	0 (0.974)	3 (0.946)	5 (0.760)	4 (0.932)	0 (0.977)	0 (0.971)
WFG 3	7	6 (0.105)	2 (0.975)	3 (0.961)	5 (0.709)	4 (0.958)	0 (0.983)	0 (0.982)
	10	6 (0.088)	1 (0.973)	3 (0.947)	5 (0.792)	4 (0.933)	0 (0.980)	1 (0.976)
	25	6 (0.037)	0 (0.983)	3 (0.965)	5 (0.758)	3 (0.963)	1 (0.974)	1 (0.977)
	50	6 (0.059)	0 (0.981)	2 (0.972)	5 (0.731)	2 (0.973)	0 (0.976)	0 (0.979)

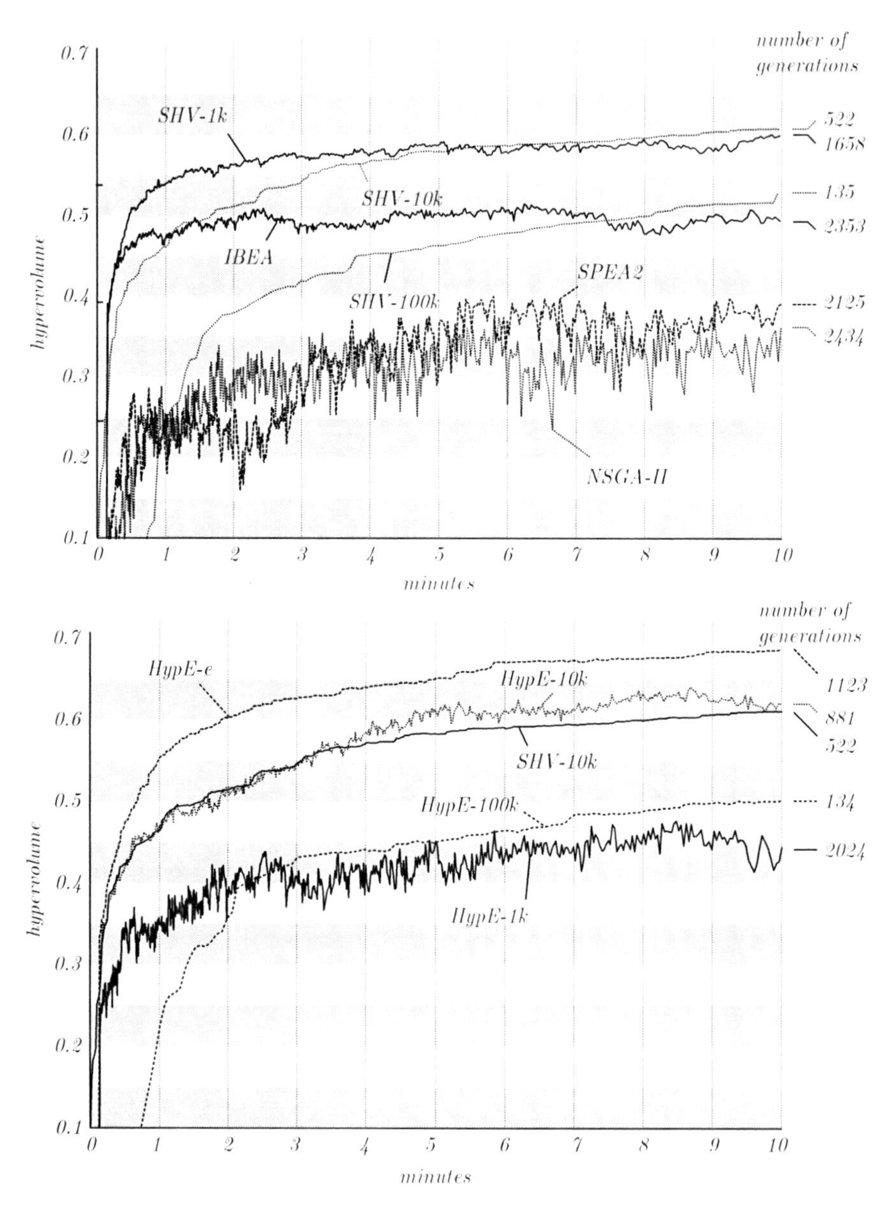

Fig. 6 Hypervolume process over 30 min of HypE and SHV for 100; 1,000; 10,000 and 100,000 samples each, denoted by the suffices 100, $1k$, $10k$ and $100k$ respectively. The test problem is WFG9 for three objectives and NSGA-II, SPEA2, and IBEA are shown as reference. The numbers at *the right border* of the figures indicate the total number of generations reached after 30 min. The results are split in two figures with identical axis for the sake of clarity

total (33), followed by HypE* (55), IBEA (55) and SHV, the method proposed in Bader et al. (2008) (97). SPEA2 and NSGA-II reach almost the same score (136 and 140 respectively), clearly outperforming random selection (217). For five out of six testproblems HypE obtains better hypervolume values than SHV. On DTLZ7 however, SHV as well as IBEA outperform HypE. This might be due to the discontinuous shape of the DTLZ7 testfunction, for which the advanced fitness scheme does not give an advantage.

The better Pareto set approximations of HypE come at the expense of longer execution time, e.g., in comparison to SPEA2 or NSGA-II. We therefore investigate, whether the fast NSGA-II and SPEA2 will not overtake HypE given a constant amount of time. Figure 6 shows the hypervolume of the Pareto set approximations over time for HypE using the exact fitness values as well as the estimated values for different samples sizes M.

Even though SPEA2, NSGA-II and even IBEA are able to process twice as many generations as the exact HypE, they do not reach its hypervolume. In the three dimensional example used, HypE can be run sufficiently fast without approximating the fitness values. Nevertheless, the sampled version is used as well to show the dependency of the execution time and quality on the number of samples M. Via M, the execution time of HypE can be traded off against the quality of the Pareto set approximation. The fewer samples are used, the more the behavior of HypE resembles random selection. On the other hand by increasing M, the quality of exact calculation can be achieved, increasing the execution time, though. For example, with $M = 1{,}000$, HypE is able to carry out nearly the same number of generations as SPEA2 or NSGA-II, but the Pareto set is just as good as when 100,000 samples are used, needing only a fifteenth the number of generations. In the example given, $M = 10{,}000$ represents the best compromise, but the number of samples should be increased in two cases: (1) when the fitness evaluation take more time, and (2) when more generations are used. The former will affect the faster algorithm much more and increasing the number of samples will influence the execution time much less. (ii) More generations are used. In the latter case, HypE using more samples might overtake the faster versions with fewer samples, since those are more vulnerable to stagnation.

In this three objective scenario, SHV can compete with HypE. This is mainly because the sampling boxes that SHV relies on are tight for small number of objectives; as this number increases, however, experiments not shown here indicate that the quality of SHV decreases in relation to HypE.

5 Conclusion

This paper proposes HypE, a novel hypervolume-based evolutionary algorithm. Both its environmental and mating selection step rely on a new fitness assignment scheme based on the Lebesgue measure, where the values can be both exactly calculated and estimated by means of sampling. In contrast to other hypervolume based algorithms, HypE is thus applicable to higher dimensional problems.

The performance of HypE was compared to other algorithms against the hypervolume indicator of the Pareto set approximation. In particular, the algorithms were tested on test problems of the WFG and DTLZ test suite. Simulations results show that HypE is a competitive search algorithm; this especially applies to higher dimensional problems, which indicates using the Lebesgue measure on many objectives is a convincing approach.

A promising direction of future research is the development of advanced adaptive sampling strategies that exploit the available computing resources most effectively.

HypE is available for download at http://www.tik.ee.ethz.ch/sop/ → download/supplementary/hype/.

Acknowledgements Johannes Bader has been supported by the Indo-Swiss Joint Research Program IT14.

References

Bader J, Zitzler E (2008) HypE: an algorithm for fast hypervolume-based many-objective optimization. TIK Report 286, Computer Engineering and Networks Laboratory (TIK), ETH Zurich, November 2008

Bader J, Deb K, Zitzler E (2008) Faster hypervolume-based search using Monte Carlo sampling. In: Conference on multiple criteria decision making (MCDM 2008). Springer, Berlin

Beume N, Rudolph G (2006) Faster S-metric calculation by considering dominated hypervolume as Klee's measure problem. Technical Report CI-216/06, Sonderforschungsbereich 531 Computational Intelligence, Universität Dortmund. Shorter version published at IASTED international conference on computational intelligence (CI 2006)

Beume N, Fonseca CM, Lopez-Ibanez M, Paquete L, Vahrenhold J (2007a) On the complexity of computing the hypervolume indicator. Technical Report CI-235/07, University of Dortmund, December 2007

Beume N, Naujoks B, Emmerich M (2007b) SMS-EMOA: multiobjective selection based on dominated hypervolume. Eur J Oper Res 181:1653–1669

Bleuler S, Laumanns M, Thiele L, Zitzler E (2003) PISA – a platform and programming language independent interface for search algorithms. In: Fonseca CM et al (eds) Conference on evolutionary multi-criterion optimization (EMO 2003). LNCS, vol 2632. Springer, Berlin, pp 494–508

Bradstreet L, Barone L, While L (2006) Maximising hypervolume for selection in multi-objective evolutionary algorithms. In: Congress on evolutionary computation (CEC 2006), pp 6208–6215, Vancouver, BC, Canada. IEEE

Bringmann K, Friedrich T (2008) Approximating the volume of unions and intersections of high-dimensional geometric objects. In: Hong SH, Nagamochi H, Fukunaga T (eds) International symposium on algorithms and computation (ISAAC 2008), LNCS, vol 5369. Springer, Berlin, pp 436–447

Brockhoff D, Zitzler E (2007) Improving hypervolume-based multiobjective evolutionary algorithms by using objective reduction methods. In: Congress on evolutionary computation (CEC 2007), pp 2086–2093. IEEE Press

Conover WJ (1999) Practical nonparametric statistics, 3rd edn. Wiley, New York

Deb K (2001) Multi-objective optimization using evolutionary algorithms. Wiley, Chichester, UK

Deb K, Agrawal S, Pratap A, Meyarivan T (2000) A fast elitist non-dominated sorting genetic algorithm for multi-objective optimization: NSGA-II. In: Schoenauer M et al (eds) Conference on parallel problem solving from nature (PPSN VI). LNCS, vol 1917. Springer, Berlin, pp. 849–858

Deb K, Thiele L, Laumanns M, Zitzler E (2005) Scalable test problems for evolutionary multiobjective optimization. In: Abraham A, Jain R, Goldberg R (eds) Evolutionary multiobjective optimization: theoretical advances and applications, chap 6. Springer, Berlin, pp 105–145

Emmerich M, Beume N, Naujoks B (2005) An EMO algorithm using the hypervolume measure as selection criterion. In: Conference on evolutionary multi-criterion optimization (EMO 2005). LNCS, vol 3410. Springer, Berlin, pp 62–76

Emmerich M, Deutz A, Beume N (2007) Gradient-based/evolutionary relay hybrid for computing Pareto front approximations maximizing the S-metric. In: Hybrid metaheuristics. Springer, Berlin, pp 140–156

Everson R, Fieldsend J, Singh S (2002) Full elite-sets for multiobjective optimisation. In: Parmee IC (ed) Conference on adaptive computing in design and manufacture (ADCM 2002), pp 343–354. Springer, London

Fleischer M (2003) The measure of Pareto optima. Applications to multi-objective metaheuristics. In: Fonseca CM et al (eds) Conference on evolutionary multi-criterion optimization (EMO 2003), Faro, Portugal. LNCS, vol 2632. Springer, Berlin, pp 519–533

Fonseca CM, Paquete L, López-Ibáñez M (2006) An improved dimension-sweep algorithm for the hypervolume indicator. In: Congress on evolutionary computation (CEC 2006), pp 1157–1163, Sheraton Vancouver Wall Centre Hotel, Vancouver, BC Canada. IEEE Press

Goldberg DE (1989) Genetic algorithms in search, optimization, and machine learning. Addison-Wesley, Reading, MA

Huband S, Hingston P, White L, Barone L (2003) An evolution strategy with probabilistic mutation for multi-objective optimisation. In: Congress on evolutionary computation (CEC 2003), vol 3, pp 2284–2291, Canberra, Australia. IEEE Press.

Huband S, Hingston P, Barone L, While L (2006) A review of multiobjective test problems and a scalable test problem toolkit. IEEE Trans Evol Comput 10(5):477–506

Igel C, Hansen N, Roth S (2007) Covariance matrix adaptation for multi-objective optimization. Evol Comput 15(1):1–28

Knowles JD (2002) Local-search and hybrid evolutionary algorithms for Pareto optimization. PhD thesis, University of Reading

Knowles J, Corne D (2003) Properties of an adaptive archiving algorithm for storing nondominated vectors. IEEE Trans Evol Comput 7(2):100–116

Laumanns M, Rudolph G, Schwefel H-P (1999) Approximating the Pareto set: concepts, diversity issues, and performance assessment. Technical Report CI-7299, University of Dortmund

Mostaghim S, Branke J, Schmeck H (2007) Multi-objective particle swarm optimization on computer grids. In: Proceedings of the 9th annual conference on genetic and evolutionary computation (GECCO 2007), pp 869–875, New York, USA. ACM

Nicolini M (2005) A two-level evolutionary approach to multi-criterion optimization of water supply systems. In: Conference on evolutionary multi-criterion optimization (EMO 2005). LNCS, vol 3410. Springer, Berlin, pp 736–751

Van Veldhuizen DA (1999) Multiobjective evolutionary algorithms: classifications, analyses, and new innovations. PhD thesis, Graduate School of Engineering, Air Force Institute of Technology, Air University

Wagner T, Beume N, Naujoks B (2007) Pareto-, aggregation-, and indicator-based methods in many-objective optimization. In: Obayashi S et al (eds) Conference on evolutionary multi-criterion optimization (EMO 2007). LNCS, vol 4403. Springer, Berlin, pp 742–756. Extended version published as internal report of Sonderforschungsbereich 531 Computational Intelligence CI-217/06, Universität Dortmund, September 2006

While L (2005) A new analysis of the LebMeasure algorithm for calculating hypervolume. In: Conference on evolutionary multi-criterion optimization (EMO 2005), Guanajuato, México. LNCS, vol 3410. Springer, Berlin, pp 326–340

While L, Hingston P, Barone L, Huband S (2006) A faster algorithm for calculating hypervolume. IEEE Trans Evol Comput 10(1):29–38

Yang Q, Ding S (2007) Novel algorithm to calculate hypervolume indicator of Pareto approximation set. In: Advanced intelligent computing theories and applications. With aspects of theoretical and methodological issues, Third international conference on intelligent computing (ICIC 2007), vol 2, pp 235–244

Zamora LP, Burguete STG (2008) Second-order preferences in group decision making. Oper Res Lett 36:99–102

Zitzler E (1999) Evolutionary algorithms for multiobjective optimization: methods and applications. PhD thesis, ETH Zurich, Switzerland

Zitzler E (2001) Hypervolume metric calculation. ftp://ftp.tik.ee.ethz.ch/pub/people/zitzler/hypervol.c

Zitzler E, Künzli S (2004) Indicator-based selection in multiobjective search. In: Yao X et al (eds) Conference on parallel problem solving from nature (PPSN VIII). LNCS, vol 3242. Springer, Berlin, pp 832–842

Zitzler E, Thiele L (1998a) An evolutionary approach for multiobjective optimization: the strength Pareto approach. TIK Report 43, Computer Engineering and Networks Laboratory (TIK), ETH Zurich

Zitzler E, Thiele L (1998b) Multiobjective optimization using evolutionary algorithms – a comparative case study. In: Conference on parallel problem solving from nature (PPSN V), pp 292–301, Amsterdam

Zitzler E, Thiele L (1999) Multiobjective evolutionary algorithms: a comparative case study and the strength Pareto approach. IEEE Trans Evol Comput 3(4):257–271

Zitzler E, Laumanns M, Thiele L (2002) SPEA2: improving the strength Pareto evolutionary algorithm for multiobjective optimization. In: Giannakoglou KC et al (eds) Evolutionary methods for design, optimisation and control with application to industrial problems (EUROGEN 2001), pp 95–100. International Center for Numerical Methods in Engineering (CIMNE)

Zitzler E, Thiele L, Laumanns M, Fonseca CM, Grunert da Fonseca V (2003) Performance assessment of multiobjective optimizers: an analysis and review. IEEE Trans Evol Comput 7(2):117–132

Zitzler E, Brockhoff D, Thiele L (2007) The hypervolume indicator revisited: on the design of Pareto-compliant indicators via weighted integration. In: Obayashi S et al (eds) Conference on evolutionary multi-criterion optimization (EMO 2007). LNCS, vol 4403. Springer, Berlin, pp 862–876

Zitzler E, Thiele L, Bader J (2008) On set-based multiobjective optimization. TIK Report 300, Computer Engineering and Networks Laboratory (TIK), ETH Zurich

Minimizing Vector Risk Measures

Alejandro Balbás, Beatriz Balbás, and Raquel Balbás

Abstract The minimization of risk functions is becoming very important due to its interesting applications in Mathematical Finance and Actuarial Mathematics. This paper addresses this issue in a general framework. Vector optimization problems involving many types of risk functions are studied. The "balance space approach" of multiobjective optimization and a general representation theorem of risk functions is used in order to transform the initial minimization problem in an equivalent one that is convex and usually linear. This new problem permits us to characterize optimality by saddle point properties that easily apply in practice. Applications in finance and insurance are presented.

1 Introduction

General risk functions are becoming very important in finance and insurance. Since the seminal paper of Artzner et al. (1999) introduced the axioms and properties of their "Coherent Measures of Risk", many authors have extended the discussion. The recent development of new markets (insurance or weather linked derivatives, commodity derivatives, energy/electricity markets, etc.) and products (inflation-linked bonds, equity indexes annuities or unit-links, hedge funds, etc.), the necessity of managing new types of risk (credit risk, operational risk, etc.) and the (often legal) obligation of providing initial capital requirements have made it rather convenient to overcome the variance as the most important risk measure and to introduce more general risk functions allowing us to address far more complex problems.

A. Balbás (✉)
University Carlos III of Madrid, CL. Madrid 126, 28903 Getafe, Madrid, Spain
e-mail: alejandro.balbas@uc3m.es

[1] It has been proved that the variance is not compatible with the Second Order Stochastic Dominance if asymmetries and/or heavy tails are involved. See Ogryczak and Ruszczynski (2002) for a very interesting analysis on the compatibility of more complex risk functions.

D. Jones et al. (eds.), *New Developments in Multiple Objective and Goal Programming*, Lecture Notes in Economics and Mathematical Systems 638,
DOI 10.1007/978-3-642-10354-4_4,

Hence, it is not surprising that the recent literature presents many interesting contributions focusing on new methods for measuring risk levels. Among others, Föllmer and Schied (2002) have defined the Convex Risk Measures, Goovaerts et al. (2004) have introduced the Consistent Risk Measures, and Rockafellar et al. (2006a) have defined the General Deviations and the Expectation Bounded Risk Measures.

Many classical actuarial and financial problems lead to optimization problems and have been revisited by using new risk functions. So, dealing with Portfolio Choice Problems, Mansini et al. (2007) use the Conditional Value at Risk ($CVaR$) and other complex risk measures in a discrete probability space, Alexander et al. (2006) compare the minimization of the Value at Risk (VaR) and the $CVaR$ for a portfolio of derivatives, Calafiore (2007) studies "robust" efficient portfolios in discrete probability spaces if risk levels are given by standard deviations or absolute deviations, and Schied (2007) deals with Optimal Investment with Convex Risk Measures.

Pricing and hedging issues in incomplete markets have also been studied (Föllmer and Schied 2002; Nakano 2004; Staum 2004; etc.) as well as Optimal Reinsurance Problems involving the $CVaR$ and stop loss reinsurance contracts (Cai and Tan 2007), and other practical problems.

Risk functions are almost never differentiable, which makes it rather difficult to provide general optimality conditions. This provokes that many authors must look for concrete properties of the special problem they are dealing with in order to find its solutions. Recent approaches by Rockafellar et al. (2006b) and Ruszczynski and Shapiro (2006) use the convexity of many risk functions so as to give general optimality conditions based on the sub-differential of the risk measure and the Fenchel Duality Theory (Luenberger 1969). The present article follows the ideas of the interesting papers above, in the sense that it strongly depends on Classical Duality Theory, but we attempt to use more properties of many risk functions that will enable us to yield new and alternative necessary and sufficient optimality conditions. Furthermore, since there is not any consensus with respect to "the best risk measure" to use in many practical applications, and the final result of many problems may critically depend on the risk measurement methodology we draw on, a second important difference between our approach and the previous literature is that we will deal with the simultaneous minimization of several risk functions, i.e., we will consider multiobjective problems. Bearing in mind the important topics of Mathematical Finance and Actuarial Mathematics that involve the minimization of risk measures, the discovery of new simple and practical rules seems to be a major objective.

The article's outline is as follows. Section 2 will present the general properties of the vector risk measure $\rho = (\rho_1, \rho_2, \ldots, \rho_r)$ and the optimization problem we are going to deal with. Since ρ is not differentiable in general, the optimization problem is not differentiable either, and Sect. 3 will be devoted to overcome this caveat. We will use the Balance Space Approach of multiobjective optimization (Balbás et al. 2005) and the Representation Theorem of Risk Measures so as to transform the initial optimization problem in an equivalent one that is differentiable and often linear. This goal is achieved by following and extending an original idea of Balbás et al. (2009).[2] However, the new problem involves new infinite dimensional Banach spaces of σ-additive measures, which provokes high degree of complexity when dealing with duality and optimality conditions. Therefore, the Mean Value Theorem (Lemma 3) is one of the most important results in this section and in the whole paper, since it will absolutely simplify the dual problem. As a consequence, Theorem 4 characterizes the optimal solutions by saddle points of a bilinear function of the feasible set and the sub-gradients of the risk measures to be simultaneously optimized. This seems to be profound finding whose proof is based on major results in Functional Analysis. Besides, the provided necessary and sufficient optimality conditions are quite different if one compares with those of previous literature. They are very general and easily apply in practical situations.

Section 4 presents two classical examples of Actuarial and Financial Mathematics that may be studied by minimizing risks. They are the Optimal Reinsurance Problem and the Portfolio Selection Problem. The novelty is given by the form of the problems, the level of generality of the analysis and the high weakness of the assumptions. The two examples are very important in practice, but this is not an exhaustive list of the real-world issues related to the optimization of risk functions. Another very interesting topics, like credit or operational risk, may be considered.

The last section of the paper points out the most important conclusions.

2 Dealing with Vector Risk Functions

Consider a probability space $(\Omega, \mathcal{F}, \mu)$ composed of the set of states of the word Ω, the σ-algebra $\mathcal{F}$ indicating the information available at a future date T, and the probability measure μ. Consider also $p \in [1, \infty)$ and $q \in (1, \infty]$ such that $1/p + 1/q = 1$, and the corresponding Banach spaces L^p and L^q. It is known that L^q is the dual space of L^p (Luenberger 1969). We will deal with a vector

$$\rho = (\rho_1, \rho_2, \ldots, \rho_r)$$

of risk functions

$$\rho_j : L^p \longrightarrow \mathbb{R}$$

such that the following condition holds:[3]

[2] Balbás and Romera (2007) also dealt with an infinite-dimensional linear optimization problem that allows us to hedge the interest rate risk, and Balbás et al. (2009) used Risk Measures Representation Theorems so as to extend the discussion and involve more general and complex sorts of risk.

[3] Hereafter $\mathbb{E}(x)$ will denote the mathematical expectation of the random variable x.

Assumption I. There exists $\varkappa_j \in \mathbb{R}$, $j = 1, 2, \ldots, r$, such that

$$\Delta^q_{(\rho_j, \varkappa_j)} = \left\{ z \in L^q; -\mathbb{E}(yz) - \varkappa_j \leq \rho_j(y) \;\; \forall y \in L^p \right\} \tag{1}$$

is $\sigma(L^q, L^p)$-compact.[4]

Proposition 1. *Fix* $j \in \{1, 2, \ldots, r\}$.

(a) The sets $\Delta^q_{(\rho_j, \varkappa_j)}$,

$$\Delta_{(\rho_j, \varkappa_j)} = \left\{ (z, k) \in L^q \times \left(-\infty, \varkappa_j\right];\; -\mathbb{E}(yz) - k \leq \rho(y) \;\; \forall y \in L^p \right\} \tag{2}$$

and

$$\Delta^{\mathbb{R}}_{(\rho_j, \varkappa_j)} = \left\{ k \in \mathbb{R};\; (z, k) \in \Delta_{(\rho_j, \varkappa_j)} \; for \; some \; z \in L^q \right\}$$

are convex. Moreover, $\Delta^q_{(\rho_j, \varkappa_j)}$ *is the natural projection of* $\Delta_{(\rho_j, \varkappa_j)}$ *on* L^q*, whereas* $\Delta^{\mathbb{R}}_{(\rho_j, \varkappa_j)}$ *is its natural projection on* $\mathbb{R}$.

(b) Under Assumption I the set $\Delta_{(\rho_j, \varkappa_j)}$ *is compact when endowed with the topology* $\tilde{\sigma}$*, product topology of* σ^* *and the usual one of the real line. Furthermore,* $\Delta^{\mathbb{R}}_{(\rho_j, \varkappa_j)}$ *is also compact and* $\Delta_{(\rho_j, \varkappa_j)}$ *is included in the* $\tilde{\sigma}$*-compact set* $\Delta^q_{(\rho_j, \varkappa_j)} \times \Delta^{\mathbb{R}}_{(\rho_j, \varkappa_j)}$.

Proof. (a) is trivial, so let us prove (b). Since the inclusion $\Delta_{(\rho_j, \varkappa_j)} \subset \Delta^q_{(\rho_j, \varkappa_j)} \times \Delta^{\mathbb{R}}_{(\rho_j, \varkappa_j)}$ is obvious it is sufficient to show that $\Delta^{\mathbb{R}}_{(\rho_j, \varkappa_j)}$ is compact and $\Delta_{(\rho_j, \varkappa_j)}$ is closed.

To see that $\Delta^{\mathbb{R}}_{(\rho_j, \varkappa_j)}$ is compact let us prove that it is closed and bounded. To see that it is closed let as assume that $(k_n)_{n \in \mathbb{N}}$ is a sequence in $\Delta^{\mathbb{R}}_{(\rho_j, \varkappa_j)}$ that converges to $k \in \mathbb{R}$. Take a sequence $(z_n, k_n)_{n \in \mathbb{N}} \subset \Delta_{(\rho_j, \varkappa_j)}$. Since $\Delta^q_{(\rho_j, \varkappa_j)}$ is compact take an agglomeration point z of $(z_n)_{n \in \mathbb{N}}$. Then it is easy to see that (z, k) is an agglomeration point of $(z_n, k_n)_{n \in \mathbb{N}}$. Thus,

$$-\mathbb{E}(yz_n) - k_n \leq \rho_j(y)$$

for every $n \in \mathbb{N}$ and every $y \in L^p$ leads to

$$-\mathbb{E}(yz) - k \leq \rho_j(y)$$

for every $y \in L^p$, and $(z, k) \in \Delta_{(\rho_j, \varkappa_j)}$, i.e., $k \in \Delta^{\mathbb{R}}_{(\rho_j, \varkappa_j)}$.

[4] In order to simplify the notation, henceforth the $\sigma(L^q, L^p)$ topology will be denoted by σ^*.

To see that $\Delta^{R}_{(\rho_j,\varkappa_j)}$ is bounded it is sufficient to prove that it is bounded from below, since $\varkappa_j$ is an obvious upper bound. Expression (2) leads to

$$-\mathbb{E}(0) - k \leq \rho_j(0),$$

for every $k \in \Delta^{R}_{(\rho_j,\varkappa_j)}$, and $\mathbb{E}(0) = 0$ implies that $k \geq -\rho_j(0)$ for every $k \in \Delta^{R}_{(\rho_j,\varkappa_j)}$.

To see that $\Delta_{(\rho_j,\varkappa_j)}$ is closed consider the net $(z_i, k_i)_{i \in I} \subset \Delta_{(\rho_j,\varkappa_j)}$ and its limit (z, k). Then,

$$-\mathbb{E}(yz_i) - k_i \leq \rho_j(y)$$

for every $i \in I$ and every $y \in L^p$ leads to

$$-\mathbb{E}(yz) - k \leq \rho_j(y)$$

for every $y \in L^p$, so $(z, k) \in \Delta_{(\rho_j,\varkappa_j)}$. □

Remark 1. As a consequence of the latter result and its proof Assumption *I* implies that $\Delta^{R}_{(\rho_j,\varkappa_j)}$ is a bounded closed interval

$$\Delta^{R}_{(\rho,\varkappa)} = \left[\varkappa_{0,j}, \varkappa_j\right] \subset \left[-\rho_j(0), \varkappa_j\right]. \tag{3}$$

Furthermore, as shown in the proof above, $\varkappa_{0,j} \geq -\rho_j(0)$.

We will also impose the following assumption:

Assumption II. The equality

$$\rho_j(y) = \text{Max}\left\{-\mathbb{E}(yz) - k;\ (z, k) \in \Delta_{(\rho_j,\varkappa_j)}\right\} \tag{4}$$

holds for every $y \in L^p$ and every $j = 1, 2, \ldots, r$.[5]

Next let us provide a proposition with a trivial (and therefore omitted) proof.

Proposition 2. *Under Assumptions I and II ρ_j is a convex function for $j = 1, 2, \ldots, r$.*

[5] Assumptions I and II frequently hold. For instance, they are always fulfilled if ρ_j is expectation bounded or a general deviation, in the sense of Rockafellar et al. (2006a) (in which case $\varkappa_{0,j} = \kappa_j = 0$), and often fulfilled if ρ_j is coherent (Artzner et al. 1999) or consistent Goovaerts et al. (2004). Furthermore, many convex risk measures (Föllmer and Schied 2002) also satisfy these assumptions.

Particular examples are the Absolute Deviation, the Standard Deviation, Down Side Semi-Demiations, the $CVaR$, the Wang Measure and the Dual Power Transform (Wang 2000, see also Cherny 2006).

Consider now a convex subset X included in an arbitrary vector space and a function

$$f : X \longrightarrow L^p$$

such that

$$\rho \circ f : X \longrightarrow \mathbb{R}^r$$

is convex. Possible examples arise when f is concave and ρ is decreasing (for instance, if every ρ_j is a coherent measure of risk) or if f is an affine function, i.e.,

$$f(tx_1 + (1-t)x_2) = tf(x_1) + (1-t) f(x_2)$$

holds for every $t \in [0, 1]$ and every $x_1, x_2 \in X$. We will deal with the multiobjective optimization problem

$$\begin{cases} \text{Min } \rho \circ f(x) \\ x \in X \end{cases}. \tag{5}$$

3 Saddle Point Optimality Conditions

Since (5) is convex, for every Pareto solution $x_0 \in X$ there exists $\alpha = (\alpha_1, \alpha_2, \ldots, \alpha_r) \geq (0, 0, \ldots, 0)$ such that $\sum_{j=1}^{r} \alpha_j = 1$ and $x_0 \in X$ solves

$$\begin{cases} \text{Min } \sum_{j=1}^{r} \alpha_j \rho_j \circ f(x) \\ x \in X \end{cases}. \tag{6}$$

The very well-known scalarization method consists in choosing an "adequate α" and then solving the problem (6) above. "Adequate α" means that α must be selected according to the decision maker preferences.

However, in this paper we will follow an alternative approach based on the notion of "Balance Point" (Galperin and Wiecek 1999 or Balbás et al. 2005, among others), since it will allow us to provide saddle point necessary and sufficient optimality conditions for (5).

So, consider that $d = (d_1, d_2, \ldots, d_r)$ is composed of strictly positive numbers and plays the role of "direction of preferential deviations" (Galperin and Wiecek 1999). Let us suppose the existence of a Pareto solution of (5) in the direction of d. According to Galperin and Wiecek (1999) d can be chosen by the decision maker depending on her/his preferences, and it indicates the marginal worsening of a given objective with respect to the improvement of an alternative one. If we assume the existence of "an ideal point" $\Upsilon \in \mathbb{R}^r$ whose coordinates are the optimal values of (5) when ρ_j substitutes ρ,[6] Balbás et al. (2005) have shown that if (x^*, θ^*) is a

[6] This assumption may be significantly relaxed (see Balbás et al. 2005), but it simplifies the exposition.

solution of the scalar problem

$$\begin{cases} \text{Min } \theta \\ \theta d + \Upsilon \geq \rho \circ f(x) \\ \theta \in \mathbb{R},\ x \in X \end{cases} \tag{7}$$

then x^* is a Pareto solution of (5) that satisfies

$$\rho \circ f\left(x^*\right) = \Upsilon + d\theta^*.$$

Conversely, for every Pareto solution x^* of (5) such that

$$\rho_j \circ f\left(x^*\right) > \Upsilon_j,$$

$j = 1, 2, \ldots, r$, there exist $d_1, d_2, \ldots, d_r > 0$ and $\theta^* > 0$ such that (x^*, θ^*) solves (7).[7]

Equation (4) clearly implies the equivalence between (7) and

$$\begin{cases} \text{Min } \theta \\ d_j\theta + \mathbb{E}\left(f(x) z_j\right) + k_j + \Upsilon_j \geq 0,\ \forall \left(z_j, k_j\right) \in \Delta_{(\rho_j, \varkappa_j)},\ j = 1, 2, \ldots, r. \\ \theta \in \mathbb{R},\ x \in X \end{cases} \tag{8}$$

The solutions of (8) will be characterized by a saddle point condition. In order to reach this result we need some additional notations and a crucial instrumental lemma. Hereafter $\mathcal{C}\left(\Delta_{(\rho_j, \varkappa_j)}\right)$, $j = 1, 2, \ldots, r$, will represent the Banach space composed of the real-valued σ^*-continuous functions on the σ^*-compact space $\Delta_{(\rho_j, \varkappa_j)}$. Similarly, $\mathcal{M}\left(\Delta_{(\rho_j, \varkappa_j)}\right)$ will denote the Banach space of σ-additive inner regular measures on the Borel σ-algebra of $\Delta_{(\rho_j, \varkappa_j)}$ (Horvàth 1966 or Luenberger 1969), and $\mathcal{P}\left(\Delta_{(\rho_j, \varkappa_j)}\right) \subset \mathcal{M}\left(\Delta_{(\rho_j, \varkappa_j)}\right)$ will be the set of inner regular probabilities. Recall that $\mathcal{M}\left(\Delta_{(\rho_j, \varkappa_j)}\right)$ is the dual space of $\mathcal{C}\left(\Delta_{(\rho_j, \varkappa_j)}\right)$.

Lemma 1 (Mean Value Theorem). *Fix $j \in \{1, 2, \ldots, r\}$. If $\nu \in \mathcal{P}\left(\Delta_{(\rho_j, \varkappa_j)}\right)$ then there exist $z_\nu \in \Delta^q_{(\rho_j, \varkappa_j)}$ and $k_\nu \in \left[\varkappa_{0,j}, \varkappa_j\right]$ such that $(z_\nu, k_\nu) \in \Delta_{(\rho_j, \varkappa_j)}$,*

$$\int_{\Delta^q_{(\rho_j, \varkappa_j)}} \mathbb{E}(yz)\, d\nu_q(z) = E\left(y z_\nu\right) \tag{9}$$

holds for every $y \in L^p$ and

[7] Moreover, this converse implication would also hold even if (5) were not a convex problem.

$$\int_{\varkappa_{0,j}}^{\varkappa_j} k d\nu_{\mathrm{R}}(k) = k_\nu. \tag{10}$$

Proof. Consider the natural projections $\nu_q \in \mathcal{P}\left(\Delta^q_{(\rho_j,\varkappa_j)}\right)$ and $\nu_{\mathrm{R}} \in \mathcal{P}\left[\varkappa_{0,j}, \varkappa_j\right]$ of ν, and the function

$$L^p \ni y \longrightarrow \psi(y) = \int_{\Delta^q_{(\rho_j,\varkappa_j)}} \mathbb{E}(yz)\, d\nu_q(z) \in \mathbb{R}.$$

It is obvious that ψ is linear so let us prove that it is also continuous. If $\Delta^q_{(\rho_j,\varkappa_j)}$ were bounded then there would exist $M \in \mathbb{R}$ such that $\|z\|_q \leq M$ for every $z \in \Delta^q_{(\rho_j,\varkappa_j)}$. Then the Hölder inequality (Luenberger 1969) would lead to

$$|\mathbb{E}(yz)| \leq \|y\|_p \|z\|_q \leq \|y\|_p M$$

for every $y \in L^p$ and every $z \in \Delta^q_{(\rho_j,\varkappa_j)}$, and

$$|\psi(y)| \leq \int M \|y\|_p\, d\nu_q(z) = M \|y\|_p$$

for every $y \in L^p$. Whence ψ would be continuous (Horvàth 1966 or Luenberger 1969). Let us see now that $\Delta^q_{(\rho_j,\varkappa_j)}$ is bounded. Since it is σ^*-compact the set $\left\{\mathbb{E}(yz); z \in \Delta^q_{(\rho_j,\varkappa_j)}\right\} \subset \mathbb{R}$ is bounded for every $y \in L^p$ because

$$L^q \ni z \longrightarrow \mathbb{E}(yz) \in \mathbb{R}$$

is σ^*-continuous. Then the Banach–Steinhaus Theorem (Horvàth 1966) shows that $\Delta^q_{(\rho_j,\varkappa_j)}$ is bounded.

Since ψ is continuous the Riesz Representation Theorem (Horvàth 1966) shows the existence of $z_\nu \in L^q$ such that (9) holds.

Besides, the inequalities

$$\varkappa_{0,j} \leq \int_{\varkappa_{0,j}}^{\varkappa_j} k d\nu_{\mathrm{R}}(k) \leq \varkappa_j$$

are obvious, so the existence of $k_\nu \in \left[\varkappa_{0,j}, \varkappa_j\right]$ satisfying (10) is obvious too.

It only remains to show that $(z_\nu, k_\nu) \in \Delta_{(\rho_j,\varkappa_j)}$. Indeed, (9) and (10) imply that

$$\begin{aligned}
-\mathbb{E}(yz_\nu) - k_\nu &= -\int_{\Delta^q_{(\rho_j,\varkappa_j)}} \mathbb{E}(yz)\,d\nu_q(z) - \int_{\varkappa_{0,j}}^{\varkappa_j} k\,d\nu_{\mathbb{R}}(k) \\
&= \int_{\Delta_{(\rho_j,\varkappa_j)}} (-\mathbb{E}(yz) - k)\,d\nu(z,k) \\
&\leq \int_{\Delta_{(\rho_j,\varkappa_j)}} \rho(y)\,d\nu(z,k) \\
&= \rho_j(y)
\end{aligned}$$

for every $y \in L^p$. □

Theorem 1 (Saddle Point Theorem). *Take $x^* \in X$ and $\theta^* \in \mathbb{R}$. (x^*, θ^*) solves (8) if and only if there exist $\left(z_j^*, k_j^*\right) \in \Delta_{(\rho_j,\varkappa_j)}$, $j = 1, 2, \ldots, r$ and*

$$\lambda^* \in \left\{ \lambda = (\lambda_1, \lambda_2, \ldots, \lambda_r);\ \sum_{j=1}^r d_j \lambda_j = 1,\ \lambda_j \geq 0,\ j = 1, 2, \ldots, r \right\}$$

such that

$$\lambda_j^* \left(d_j \theta^* + \Upsilon_j + \mathbb{E}\left(f\left(x^*\right) z_j^*\right) + k_j^*\right) = 0,$$

$j = 1, 2, \ldots, r$, and

$$\sum_{j=1}^r \lambda_j^* \left(\mathbb{E}\left(f\left(x^*\right) z_j^*\right) + k_j^*\right) \geq \sum_{j=1}^r \lambda_j^* \left(\mathbb{E}\left(f(x) z_j^*\right) + k_j^*\right) \tag{11}$$

for every $x \in X$. If so,

$$\rho_j \circ f\left(x^*\right) = -\left(k_j^* + \mathbb{E}\left(f\left(x^*\right) z_j^*\right)\right)$$

holds for every $j = 1, 2, \ldots, r$, and

$$\lambda_j^* \left(k_j^* + \mathbb{E}\left(f\left(x^*\right) z_j^*\right)\right) \leq \lambda_j^* \left(k_j + \mathbb{E}\left(f\left(x^*\right) z_j\right)\right) \tag{12}$$

holds for every $j = 1, 2, \ldots, r$ and every $\left(z_j, k_j\right) \in \Delta_{(\rho_j,\varkappa_j)}$.[8]

---

[8] Notice that (11) and (12) show that

$$\left(x^*, \left(z_j^*, k_j^*\right)\right)$$

is a Saddle Point of the function

$$X \times \Pi_{j=1}^r \Delta_{(\rho_j,\varkappa_j)} \ni \left(x, \left(z_j, k_j\right)\right) \longrightarrow \sum_{j=1}^r \lambda_j^* \left(\mathbb{E}\left(f(x) z_j\right) + k_j\right) \in \mathbb{R}.$$

Proof. The constraints of (8) are valued on the Banach space $\mathcal{C}\left(\Delta_{(\rho_j,\varkappa_j)}\right)$, $j = 1,2,\ldots,r$. Accordingly, the Lagrangian function

$$\mathcal{L} : X \times \mathbb{R} \times \prod_{j=1}^{r} \mathcal{M}\left(\Delta_{(\rho_j,\varkappa_j)}\right) \longrightarrow \mathbb{R}$$

of (8) becomes (Luenberger 1969)

$$\begin{aligned}\mathcal{L}\left(x,\theta,(\nu_j)_{j=1}^{r}\right) = {} & \theta\left(1-\sum_{j=1}^{r} d_j \int_{\Delta_{(\rho_j,\varkappa_j)}} d\nu_j\left(z_j,k_j\right)\right)\\ & -\sum_{j=1}^{r}\int_{\Delta_{(\rho_j,\varkappa_j)}} \mathbb{E}\left(f(x)\,z_j\right) d\nu_j\left(z_j,k_j\right)\\ & -\sum_{j=1}^{r}\int_{\Delta_{(\rho_j,\varkappa_j)}} k_j d\nu_j\left(z_j,k_j\right)\\ & -\sum_{j=1}^{r}\Upsilon_j\int_{\Delta_{(\rho_j,\varkappa_j)}} d\nu_j\left(z_j,k_j\right),\end{aligned}$$

that may simplify to

$$\begin{aligned}\mathcal{L}\left(x,\theta,(\nu_j)_{j=1}^{r}\right) = {} & \theta\left(1-\sum_{j=1}^{r} d_j\lambda_j\right)\\ & -\sum_{j=1}^{r}\int_{\Delta_{(\rho_j,\varkappa_j)}} \mathbb{E}\left(f(x)\,z_j\right) d\nu_j\left(z_j,k_j\right)\\ & -\sum_{j=1}^{r}\int_{\Delta_{(\rho_j,\varkappa_j)}} k_j d\nu_j\left(z_j,k_j\right)\\ & -\sum_{j=1}^{r}\Upsilon_j\lambda_j,\end{aligned}$$

if $\lambda_j = \int_{\Delta_{(\rho,\varkappa)}} d\nu_j\left(z_j,k_j\right) \geq 0$ for $j = 1,2,\ldots,r$. It is clear that the infimum

$$Inf\ \left\{\mathcal{L}\left(x,\theta,(\nu_j)_{j=1}^{r}\right) : \theta\in\mathbb{R},\ x\in X\right\} \tag{13}$$

can only be finite if $\sum_{j=1}^{r} d_j\lambda_j = 1$. Thus, the dual problem of (8), given by (13), becomes (Luenberger 1969)

$$\begin{cases} \text{Max } -\sum_{j=1}^{r} \Upsilon_j \lambda_j \\ \quad + \left(Inf_{x \in X} \left\{ -\sum_{j=1}^{r} \int_{\Delta_{(\rho_j, \varkappa_j)}} \left(\mathbb{E}\left(f(x) z_j \right) + k_j \right) d\nu_j \left(z_j, k_j \right) \right\} \right) \\ \lambda_j = \int_{\Delta_{(\rho_j, \varkappa_j)}} d\nu_j \left(z_j, k_j \right), \ j = 1, 2, \ldots, r \\ \sum_{j=1}^{r} d_j \lambda_j = 1 \\ \nu_j \geq 0, \ j = 1, 2, \ldots, r \end{cases} . \tag{14}$$

$d_j > 0$, $j = 1, 2, \ldots, r$ implies that (8) satisfies the Slater Qualification,[9] so, if (8) (or (7)) is bounded, then the dual problem above is solvable and there is no duality gap (the optimal values of (8) and (14) coincide) (Luenberger 1969). If (7) were unbounded then taking a feasible solution (x, θ) with $\theta < 0$ we would have $\Upsilon \geq \rho \circ f(x) - \theta d > \rho \circ f(x)$, against the election of Υ.

Take $\nu^* = \left(\nu_j^* \right)_{j=1}^{r}$ solving (14) and $\lambda_j^* = \nu_j^* \left(\Delta_{(\rho_j, \varkappa_j)} \right)$, $j = 1, 2, \ldots, r$. Take $\left(z_j^*, k_j^* \right) \in \Delta_{(\rho_j, \varkappa_j)}$, $j = 1, 2, \ldots, r$ satisfying the conditions of the Mean Value Theorem (previous lemma) for

$$\tilde{\nu}_j^* = \frac{\nu_j^*}{\lambda_j^*}$$

if $\lambda_j^* > 0$, and $\left(z_j^*, k_j^* \right) \in \Delta_{(\rho_j, \varkappa_j)}$, if $\lambda_j^* = 0$. According to Luenberger (1969), a (8)-feasible element (x^*, θ^*) solves (8) if and only if

$$-\sum_{j=1}^{r} \lambda_j^* \left(\mathbb{E}\left(f(x^*) z_j^* \right) + k_j^* \right) \leq -\sum_{j=1}^{r} \lambda_j^* \left(\mathbb{E}\left(f(x) z_j^* \right) + k_j^* \right)$$

for $j = 1, 2, \ldots, r$ and every $x \in X$, and

$$\lambda_j^* \left(d_j \theta^* + \mathbb{E}\left(f(x^*) z_j^* \right) + k_j^* + \Upsilon_j \right) = 0, \ j = 1, 2, \ldots, r.$$

Then, if $\lambda_j^* \neq 0$, bearing in mind the constraint of (7) we have

$$\rho_j \circ f\left(x^*\right) \leq \theta^* d_j + \Upsilon_j = -\mathbb{E}\left(f\left(x^*\right) z_j^*\right) - k_j^*,$$

so

$$-\left(k_j^* + \mathbb{E}\left(f\left(x^*\right) z_j^*\right)\right) \geq \rho_j \circ f\left(x^*\right) \geq -\left(k_j + \mathbb{E}\left(f\left(x^*\right) z_j\right)\right)$$

for every $\left(z_j, k_j\right) \in \Delta_{(\rho_j, \varkappa_j)}$, holds from the definition of $\Delta_{(\rho_j, \varkappa_j)}$. □

[9] That is, there is a least one feasible solution of (8) satisfying all the constraints in terms of strict inequalities. Indeed, $d_j > 0$, $j = 1, 2, \ldots, r$ implies that one only have to take a value of θ large enough.

4 Applications

This section is devoted to present two practical applications. The first one may be considered as "classical" in Financial Mathematics, while the second one is "classical" in Actuarial Mathematics. Both lead to optimization problems that perfectly fit on (5), so the theory above absolutely applies. The two examples are very important in practice, but this is far of being an exhaustive list of the real-world issues related to the optimization of risk functions. Another very interesting topics, like pricing and hedging issues, credit risk or operational risk, etc., may be considered.

4.1 *Portfolio Choice*

The optimal portfolio selection is probably the most famous multiobjective optimization problem in finance. Let us assume that

$$y_1, y_2, \ldots, y_n \in L^p$$

represent the random returns of n available assets,[10] and denote by $x = (x_1, x_2, \ldots, x_n) \in \mathbb{R}^n$ the portfolio composed of the percentages invested in these assets. If ρ is the ($\mathbb{R}^r$-valued) vector risk function used by the investor then he/she will select that strategy solving

$$\begin{cases} \text{Min } \rho\left(\sum_{i=1}^n x_i y_i\right) \\ \sum_{i=1}^n x_i = 1 \\ \sum_{i=1}^n x_i \mathbb{E}(y_i) \geq r_0 \end{cases} \tag{15}$$

$r_0 \in \mathbb{R}$ denoting the minimum required expected return. If some short-sale restrictions must be imposed then constraints such as $x_i \geq 0$ for some (or all) subscripts must be added. Similarly, additional equality or inequality constraints reflecting several market-linked or agent-linked restrictions may arise. It is obvious that (15) is a particular case of (5).

4.2 *Optimal Reinsurance*

The "Optimal Reinsurance Problem" is classical in Actuarial Mathematics. Many authors have dealt with it by using different "Premium Principles", and a quite general approach may be found in Kaluszka (2005), where the author uses even some

[10] That is, y_i will be the final pay-off received at a future date $t = T$ if one invests one dollar in the ith-security at the initial date $t = 0$.

coherent measures of risk to price the insurance. However, the minimized risk functions are usually classical deviations (standard deviation or absolute deviation) or classical down side semi-deviations. More recently Cai and Tan (2007) minimize the *Value at Risk* and the *Conditional Tail Expectation* (Artzner et al. 1999) for a very particular case, since they only deal with the *Expected Value Principle* and, more importantly, *stop–loss* reinsurance contracts. We will show below that the general approach of this paper may apply to minimize general risk functions in the optimal reinsurance problem and we do not need to be constrained by any kind of reinsurance contract.

Consider that an insurance company receives the fixed amount S_0 (premium) and will have to pay the random variable $y_0 \in L^p$ within a given period $[0, T]$ (claims). Suppose also that a reinsurance contract is signed in such a way that the company will only pay $x \in L^p$ whereas the reinsurer will pay $y_0 - x$. If the reinsurer premium is given by the convex function,[11]

$$\pi : L^p \longrightarrow \mathbb{R}$$

and π_1 is the highest amount that the insurer would like to pay for the contract, then the insurance company will chose x (optimal retention) so as to solve

$$\begin{cases} \operatorname{Min} \rho\left(S_0 - x - \pi\left(y_0 - x\right)\right) \\ \pi\left(y_0 - x\right) \leq \pi_1 \\ 0 \leq x \leq y_0 \end{cases} \tag{16}$$

ρ being a vector risk function. Notice that

$$x \longrightarrow S - x - \pi\left(y_0 - x\right)$$

is a concave function, so (16) is included in (5) and the developed theory obviously applies.

5 Conclusions

The minimization of risk functions is becoming very important in Mathematical Programming, Mathematical Finance and Actuarial Mathematics, which provokes a growing interest in this topic that is becoming the focus of many researchers.

Since risk functions are not differentiable there are significant difficulties when they are involved in minimization problems. Convex programming and duality

[11] Insurance premiums are usually given by convex functions. See for instance Deprez and Gerber (1985).

methods have been proposed. This paper has also followed this line of research, though there are two major differences. On the one hand, we deal with multiobjective problems, which is far more realistic due to the lack of consensus with respect to the risk function to be used in many applications. Secondly, the provided necessary and sufficient optimality conditions are quite different if one compares with previous literature. Indeed, they are related to saddle point properties of a bilinear function of the feasible set and the sub-gradient of the risk measures to be optimized. This seems to be profound finding whose proof is based on the $weak^*$-compactness of the sub-gradient of the risk measure, the duality theory in general Banach spaces and a given Mean Value Theorem for risk measures. The yielded optimality conditions easily apply in practice. Interesting applications in finance and insurance have been given.

Acknowledgements Research partially developed during the sabbatical visit to Concordia University (Montreal, Quebec, Canada). The authors thank the Department of Mathematics and Statistics for its great hospitality.

Research partially supported by "Welzia Management SGIIC SA", "RD Sistemas SA", "Comunidad Autónoma de Madrid" (Spain), Grant s-0505/tic/000230, and "MEyC" (Spain), Grant SEJ2006-15401-C04. The usual caveat applies.

References

Alexander S, Coleman TF, Li Y (2006) Minimizing $CVaR$ and VaR for a portfolio of derivatives. J Bank Finance 30:538–605

Artzner P, Delbaen F, Eber JM, Heath D (1999) Coherent measures of risk. Math Finance 9:203–228

Balbás A, Balbás R, Mayoral S (2009) Portfolio choice problems and optimal hedging with general risk functions: a simplex-like algorithm. Eur J Oper Res 192(2):603–620

Balbás A, Galperin E, Guerra PJ (2005) Sensitivity of Pareto solutions in multiobjective optimization. J Optim Theor Appl 126(2):247–264

Balbás A, Romera R (2007) Hedging bond portfolios by optimization in Banach spaces. J Optim Theor Appl 132(1):175–191

Cai J, Tan KT (2007) Optimal retention for a stop loss reinsurance under the VaR and CTE risk measures. ASTIN Bull 37(1):93–112

Calafiore GC (2007) Ambiguous risk measures and optimal robust portfolios. SIAM J Optim 18(3):853–877

Cherny AS (2006) Weighted $V@R$ and its properties. Finance Stochastics 10:367–393

Deprez O, Gerber U (1985) On convex principles of premium calculation. Insur Math Econ 4:179–189

Föllmer H, Schied A (2002) Convex measures of risk and trading constraints. Finance Stochastics 6:429–447

Galperin EA, Wiecek MM (1999) Retrieval and use of the balance set in multiobjective global optimization. Comput Math Appl 37:111–123

Goovaerts M, Kaas R, Dhaene J, Tang Q (2004) A new classes of consistent risk measures. Insur Math Econ 34:505–516

Horvàth J (1966) Topological vector spaces and distributions, vol I. Addison Wesley, Reading, MA

Kaluszka M (2005) Optimal reinsurance under convex principles of premium calculation. Insur Math Econ 36:375–398

Luenberger DG (1969) Optimization by vector spaces methods. Wiley, New York

Mansini R, Ogryczak W, Speranza MG (2007) Conditional value at risk and related linear programming models for portfolio optimization. Ann Oper Res 152:227–256
Nakano Y (2004) Efficient hedging with coherent risk measure. J Math Anal Appl 293:345–354
Ogryczak W, Ruszczynski A (2002) Dual stochastic dominance and related mean risk models. SIAM J Optim 13:60–78
Rockafellar RT, Uryasev S, Zabarankin M (2006a) Generalized deviations in risk analysis. Finance Stochastics 10:51–74
Rockafellar RT, Uryasev S, Zabarankin M (2006b) Optimality conditions in portfolio analysis with general deviations measures. Math Program B 108:515–540
Ruszczynski A, Shapiro A (2006) Optimization of convex risk functions. Math Oper Res 31(3):433–452
Schied A (2007) Optimal investments for risk- and ambiguity-averse preferences: a duality approach. Finance Stochastics 11:107–129
Staum J (2004) Fundamental theorems of asset pricing for good deal bounds. Math Finance 14:141–161
Wang SS (2000) A class of distortion operators for pricing financial and insurance risks. J Risk Insur 67:15–36

Multicriteria Programming Approach to Development Project Design with an Output Goal and a Sustainability Goal*

Enrique Ballestero

Abstract Especially intended for managers faced with development project design, this paper proposes a multicriteria decision making (MCDM) model with two objectives, output maximization and sustainability to be attained as much as possible. This proposal is motivated because: (a) aggressive managers seek to optimize the project output by using deterministic methods such as capital budgeting techniques and mathematical programming models; (b) conservative managers seek sustainability by using replicas of running development projects, which have proven reliable in practice; (c) few managers use stochastic models to ensure sustainability as these models require unavailable information on random variables and complex feedback. Then, this proposal is to articulate the aggressive output standpoint and the conservative replica standpoint into a two-objective programming model looking for a compromise solution between both goals. A numerical example on farm project design is developed in detail and discussed.

1 Introduction

Development projects are often designed as replicas of currently running projects which inspire confidence concerning sustainability. This is viewed as a practical way of sustainable design easier and less cumbersome than the design attempts based on stochastic models such as classical expected utility maximization (Von Neumann and Morgenstern 1947; Arrow 1965). Consequently, the proposition in this paper does not deal with risk and stochastic aspects at all. In contrast, the proposition establishes an objective of safety and sustainability, this objective being articulated into a multiobjective model by considering replicas of a reliable pattern.

E. Ballestero
Escuela Politécnica Superior de Alcoy, 03801 Alcoy, Alicante, Spain
e-mail: eballe@esp.upv.es

*8th International Conference on Multiple Objective and Goal Programming.

D. Jones et al. (eds.), *New Developments in Multiple Objective and Goal Programming*, Lecture Notes in Economics and Mathematical Systems 638,
DOI 10.1007/978-3-642-10354-4_5,

First, project managers have often difficulties in finding statistical data to measure risk by variance and covariance matrices (see Ballestero 2006). In other decision making scenarios such as for example portfolio choice, the analyst can use time series of prices and returns from daily observations on the markets, these data being precise and reliable. In contrast, time series are not generally available for development project design. Managers can analyze some real world cases (either historical or current in nature) related to their projects but nothing else. Available case studies on development projects hardly provide time series concerning all critical variables involved in each project. On the other hand, many project managers are not used to handling stochastic models in the framework of design.

This paper proposes an approach to development project design relying on multiobjective programming. Objectives in this model are as follows:

1. To maximize the project's output when potential impacts on sustainability are ignored, namely, when adverse consequences resulting from risk/uncertainty inherent in the development project are overlooked. This maximization objective is often considered by aggressive managers who use capital budgeting techniques and deterministic programming models in the certainty context.
2. To achieve sustainability/safety as much as possible by designing the project as a replica of a running development project which has proven reliable. This objective is often considered by conservative managers who devise a development project as a replica of another existing project, because the existing pattern is viewed as a high standard of sustainability/safety.

Indeed, sustainability of the project means that the investment can be maintained at a satisfactory operational level now and in the future up to a reasonable time horizon.

Potential users of the paper are development project managers and financial consultants especially interested in design.

Our proposal is relevant as development project design is relevant. Also, the replica-based statement is new. In fact, no similar approaches where replicas of development projects play a critical role combined with deterministic optimization are currently found in the decision making literature. Therefore, references to pave the way for reading the paper should be few. Concerning multiobjective programming in general, suitable references are Caballero et al. (1997) and Steuer (2001). Regarding sustainable clean development projects to be selected by multiple criteria approaches, see Lenzen et al. (2007). Concerning trust in construction projects, an empirical analysis is Khalfan et al. (2007).

2 Methodology

This chapter deals with development project design by multiobjective programming. Consider a development project P to be drawn up. Let $(q_1, q_2, \ldots, q_i, \ldots, q_n)$ be the vector describing the characteristics of this project. These characteristics are viewed as investments of particular resources in the project. Therefore, q_i is quantity of the ith resource to be invested. Let $(q_1^*, q_2^*, \ldots, q_i^*, \ldots, q_n^*)$ be the vector describing the characteristics of the running development project P^*, which is taken by the project manager as a pattern of sustainable design. Then, the decision variables are defined as follows:

$$x_i = q_i / q_i^*; i = 1, 2, \ldots, n \tag{1}$$

Decision variables (1) define a replica of pattern P^*.

For technical reasons, some characteristics can be excluded from the vector as they require values fixed a priori by the project manager.

2.1 *First Objective: Output Maximization*

In Chap. 1, Sect. 1, this objective is stated by maximizing a capital budgeting measure (for example, Net Present Value) as output achievement. This maximization is often formulated by linear programming (LP) or another mathematical programming model. This means that risk/uncertainty is not considered. In other words, all the data are then viewed as nonrandom variables although they were random in nature. An advantage of LP is simplicity; however, the random negative impacts on the project's safety/sustainability are disregarded. Accepting this limitation, the project manager states the following LP model:

$$\max \sum_{i=1}^{n} r_i q_i = \max \sum_{i=1}^{n} r_i q_i^* x_i \tag{2}$$

subject to

$$\sum_{i=1}^{n} a_{ij} q_i^* x_i \leq c_j; j = 1, 2, \ldots, m \tag{3}$$

together with the non-negativity conditions. As the characteristics have the meaning of particular resources to be invested in the project, each $r_i q_i$ in (2) represents a capital budgeting measure such as Net Present Value. Sometimes, the project manager can estimate these measures in terms of relative importance by using indexes r_i of benefit. In economics and management science, benefits are defined as the positive contribution to an economic value from an entrepreneurial activity or project. In any case, the r_i benefit indexes should be expressed in appropriate units for aggregation.

Constraints (3) are referred to limited resources (physical, financial, environmental, etc.). Thus, they are classified in technical constraints and budget constraints. Each coefficient a_{ij} measures a cost per unit associated with the ith characteristic in the context of the jth constraint. Parameter c_j also means a cost in the jth context.

Solution to deterministic linear program (2) to (3) is denoted as: $x_a = (x_{1a}, x_{2a}, \ldots, x_{ia}, \ldots, x_{na})$.

2.2 Second Objective: Sustainability

In Chap. 1, Sect. 2, this objective involves designing development project P as an exact or scale replica of an existing reliable (sustainable) development project, namely, the P^* pattern. This pattern is regarded by the project manager as a real world model of sustainability. Therefore, the project manager's second goal is to produce a counterpart of development project P^* that resembles the pattern as closely as possible. Accordingly, the second objective is stated as follows: to achieve either an exact or scale replica $x_b = (x_{1b}, x_{2b}, \ldots, x_{ib}, \ldots, x_{nb})$ on a scale smaller, equal or larger than the P^* original, namely,

$$x_{1b} = x_{2b} = \ldots = x_{ib} = \ldots = x_{nb} = \chi \tag{4}$$

where χ is a positive parameter. In the special case $\chi = 1$, we would have an exact replica.

Once the pattern has been chosen, the constraint system (3) together with the scale replica condition (4) is solved leading to the x_b replica vector. Notice that this vector solution lies on the frontier given by the constraint system (3).

2.3 Compromise Solution

The next purpose is to obtain a 'satisficing' compromise solution to the two-objective problem formulated from the first and the second objectives. In other words, the project manager looks for a compromise between output maximization (unrelated to the pattern) and the replica design, so that both play their role in the development project design. This suggests the following:

Assumption 1. *Compromise solution. This is the frontier point where the line defined by the following vector:*

$$x_c = \alpha x_a + \beta x_b; \alpha, \beta \geq 0, \alpha + \beta = 1 \tag{5}$$

intercepts frontier (3).

Meaning. Equation (5) means a compromise between x_a and x_b, namely, between solutions to the first and the second objective, respectively. Weights α and β of the convex combination are decided by the project manager to reflect preferences for output maximization (unrelated to the pattern) and for the replica design, respectively.

From (3) to (4), we get:

$$\chi \leq c_j \Big/ \sum_{i=1}^{n} a_{ij} q_i^*; j = 1, 2, \ldots, m \tag{6}$$

To determine vector x_b by (4), we specify χ as the lowest right hand side value in the set of constraints (6). By introducing this χ value into (4), vector x_b is determined as a frontier point. On the other hand, the x_a vector is the standard solution to the LP problem (2) to (3). Now, vectors x_a and x_b just obtained are introduced into (5), together with weights α and β previously established from the manager's preferences. Thus, vector x_c is determined.

Finally, we should determine the frontier point $x_f = (x_{1f}, x_{2f}, \ldots, x_{if}, \ldots, x_{nf})$, where vector x_c intercepts frontier (3). This final solution is given by:

$$x_f = \lambda x_c \tag{7}$$

where parameter λ is obtained by:

$$\max \lambda \tag{8}$$

subject to

$$\lambda \sum_{i=1}^{n} a_{ij} q_i^* (\alpha x_{ia} + \beta x_{ib}) \leq c_j; \; j = 1, 2, \ldots, m; \lambda \geq 0. \tag{9}$$

2.4 Feedback

Weights α and β in (5) can be modified from their initial values, thus obtaining new solutions to be evaluated in terms of output achievement and resemblance to the pattern. Suppose α and β change to $\alpha' = (1 + \varepsilon)\alpha$ and $\beta' = (1 - \alpha - \varepsilon\alpha)$, respectively, other things being equal. This change in weights leads to a new vector x_c', the difference between both the old and the new vector being:

$$x_c{}' - x_c = \varepsilon\alpha \, (x_a - x_b)$$

Therefore, the difference between both vectors tends to zero either if ε tends to zero, or if $(x_a - x_b)$ tends to zero, other things being equal. Usually in sensitivity analysis, ε is small, and $\varepsilon\alpha$ is still smaller than ε since $\alpha < 1$. In Sect. 4, further research on this issue from empirical information is foreseen.

3 An Illustrative Example

A fictitious case of development project design in agriculture is here presented, with numerical values taken from unpublished discussion notes used by the author. This is to decide farm areas of meadows, dry farming and orchards. To make their decision, the project managers attempt to imitate to a certain extent an already running agricultural project which has proven reliable and sustainable. In other words, this pattern is technically, economically and environmentally considered as an example of sustainability. Pattern P^* has the following characteristics: $q_1^* = 255.57$ ha of meadows; $q_2^* = 125.35$ ha of dry farming; and $q_3^* = 273.83$ ha of orchards.

Objectives (as defined in Chap. 1) are as follows.

3.1 First Objective: Output Maximization (Unrelated to the Pattern)

This is the solution to deterministic LP (2) to (3). Benefit indexes $r_1 = 8.30$, $r_2 = 5.93$, and $r_3 = 9.49$ in objective function (2) are estimated by the project managers from a capital budgeting perspective. Each index is expressed in monetary units per hectare. Therefore, the following LP model is formulated.

$$\max(8.30 * 255.57 * x_1 + 5.93 * 125.35 * x_2 + 9.49 * 273.83 * x_3)$$

subject to [see (3)]

- Land constraint:
  $255.57 * x_1 + 125.35 * x_2 + 273.83 * x_3 \leq 772.80$ size units.
- Investment costs:
  $67.05 * 255.57 * x_1 + 53.18 * 125.35 * x_2 + 35.84 * 273.83 * x_3 \leq 41924.19$ monetary units.
- Environmental constraint:
  $273.83 * x_3 \leq 282.34$ size units.

Non-negativity conditions: $x_1 \geq 0, x_2 \geq 0, x_3 \geq 0$.

By solving this LP model, we obtain:

$$x_a = (1.856,\ 0,\ 1.031)$$

Hence, output maximization (first objective) yields the following solution.

- Meadows: 1.856*255.57 = 474.34 ha
- Dry farming: 0*125.35 = 0 ha
- Orchards: 1.031*273.83 = 282.32 ha

3.2 Second Objective: Sustainability (Related to the Pattern)

By applying (6) to the numerical data, we get:

$$\chi \leq 772.80/(255.57 + 125.35 + 273.83) = 772.80/654.75 = 1.18$$

$$\chi \leq 41924.19/(67.05 * 255.57 + 53.18 * 125.35 + 35.84 * 273.83)$$
$$= 41924.19/33616.15 = 1.25$$
$$\chi \leq 282.34/273.83 = 1.031$$

Therefore, $\chi = 1.031$. From chain (4), we have:

$$x_b = (1.031, 1.031, 1.031).$$

Hence, the second objective yields the following sizes:

- Meadows: 1.031 * 255.57 = 263.49 ha
- Dry farming: 1.031 * 125.35 = 129.24 ha
- Orchards: 1.031 * 273.83 = 282.32 ha

3.3 Compromise Solution and Final Solution on the Frontier

They require the following tasks.

First step. Establish weights α and β according to the project manager's preferences for output maximization (first objective) and for replica (second objective). In our example, $\alpha = 0.45$ and $\beta = 0.55$, namely, the project manager slightly prefers the sustainability objective to the output maximization objective.

Second step. Compute the x_c compromise vector. From (5) with the numerical expressions of x_a and x_b obtained in Sects. 3.1 and 3.2, respectively, we get:

$$x_c = 0.45*(1.856, 0, 1.031)+0.55*(1.031, 1.031, 1.031) = (1.402, 0.567, 1.031)$$

Third step. Compute parameter λ by the auxiliary LP model (8) to (9), namely:
max λ
subject to

$$\lambda * (255.57 * 1.402 + 125.35 * 0.567 + 273.83 * 1.031) = 711.70\lambda \leq 772.80$$
$$\lambda * (67.05 * 255.57 * 1.402 + 53.18 * 125.35 * 0.567 + 35.84 * 273.83 * 1.031)$$
$$\leq 41924.19$$
$$\lambda * 273.83 * 1.031 \leq 282.34$$

together with the non-negativity condition $\lambda \geq 0$. By solving this auxiliary model, we get $\lambda = 1.000086 \approx 1$.

Fourth step. Determine the final solution x_f from (7) on the frontier, namely:

$$x_f = (1 * 1.402, 1 * 0.567, 1 * 1.031) = (1.402, 0.567, 1.031)$$

Therefore, development project P is sized as follows:

- Meadows: $1.402 * 255.57 = 358.31$ ha
- Dry farming: 0.567*125.35 = 71.07 ha
- Orchards: 1.031*273.83 = 282.32 ha

3.4 *Comparison of Results*

Let us compare the sizes given by the final x_f solution to the sizes given by the x_a and x_b solutions.

- Meadows: 24.45% smaller than the respective x_a result. Moreover, 26.47% larger than the respective x_b result.
- Dry farming: This increases from zero (in vector x_a) to 0.567 (in vector x_f). Moreover, 45% smaller than the respective x_b result.
- Orchards: here, all three solutions coincide. This is because: (a) the sizes given by the x_a and x_b solutions are both equal to 231.99; then, the compromise value between LP and the replica is also 231.99; and (b) the compromise point is brought to the frontier by the factor $\lambda = 1.000086 \approx 1$, so that the compromise value does not increase.

In short, we have:

(a) With the x_f solution, the extremely unbalanced results given by LP are avoided. This occurs with dry farming. While the LP solution was no dry farming, this abrupt result is substantially mitigated in the x_f solution. Also, meadows is reduced by around 25% over the LP result, thus correcting the too large size resulting from LP.
(b) However, the x_f solution allows the project manager to propose an original design to a certain extent. Indeed, project P has turned out to be far from being an exact replica of the pattern.

4 Concluding Remarks

A major result has been to articulate aggressiveness and conservatism in development project design. Aggressiveness involves maximizing the project's output, while conservatism involves replica projects from the axiom: "if the pattern has proven sustainable, then its replica will be probably sustainable". Solutions obtained from the two-objective programming model depend on the α/β preference ratio. Preference weights α and β have a clear meaning to the manager and they are straightforwardly elicited as occurs in decision making approaches whenever the number of weights does not exceed two. Moreover, the initial solution can be evaluated and modified by feedback –an appealing procedure to managers. By moving parameters α and β in (5), the project manager can analyze tradeoffs between output achievement and resemblance to the pattern. This allows the project manager to adjust the solution to convenient output levels or resemblance. In any case, the final solution (vector x_f) lies on the frontier of constraints (3). Certainly, the two-objective programming model developed above is not the only possible way of addressing the aggressiveness versus conservatism dilemma in development project design. A goal programming model with similar scope and purpose can be also proposed. In short, the paper has shown how development project design can be deterministically addressed (without difficult stochastic treatment) in terms of output optimization and sustainability by the replica-based approach.

Further research can be conducted on the following issues:

(a) To develop real world case studies, where managers really proposes different (α, β) weights leading to different results, which are discussed through sensitivity analysis.
(b) To introduce utility functions related to compromise programming (CP). For this purpose, a theorem connecting bi-attribute utility and CP could be applied in the above two-criterion framework (Ballestero and Romero 1998, Chap. 6).
(c) To extend the proposition in such a way, that instead of existing object, a fictitious reference object created on the basis of some existing (or fictitious) objects could be taken into consideration. This extension is suggested to the author by an anonymous referee.

Acknowledgements Thanks are given to an anonymous referee for their suggestions to improve the paper.

References

Arrow K (1965) Aspects of the theory of risk bearing. Academic Book Store, Helsinki

Ballestero E (2006) Stochastic linear programming with scarce information: an approach from expected utility and bounded rationality applied to the textile industry. Eng Optim 38(4): 425–440

Ballestero E, Romero C (1998) Multiple criteria decision making and its applications to economic problems. Kluwer, Boston

Caballero R, Ruiz F, Steuer RE (1997) Advances in multiple objective and goal programming. Lecture notes in economics and mathematical systems, vol 455. Springer, Berlin

Khalfan MMA, McDermott P, Swan W (2007) Building trust in construction projects. Supply Chain Manag Int J 12:385–391

Lenzen M, Schaeffer R, Matsuhashi R (2007) Selecting and assessing sustainable CDM projects using multi-criteria methods. Clim Pol 7(2):121–138

Steuer RE (2001) Multiobjective programming. In: Gass SI, Harris CM (eds) International encyclopedia of operations research and management science. Kluwer Academic, Boston, pp 532–538

Von Neumann J, Morgenstern O (1947) Theory of games and economic behaviour. Princeton University Press, Princeton

Automated Aggregation and Omission of Objectives for Tackling Many-Objective Problems

Dimo Brockhoff and Eckart Zitzler

Abstract Many-objective problems pose various challenges in terms of decision making and search. One approach to tackle the resulting problems is the automatic reduction of the number of objectives such that the information loss is minimized. While in a previous work we have investigated the issue of omitting objectives, we here address the generalized problem of aggregating objectives using weighted sums. To this end, heuristics are presented that iteratively remove two objectives and replace them by a new objective representing an optimally weighted combination of them. As shown in the paper, the new reduction method can substantially reduce the information loss and thereby can be highly useful when analyzing trade-off sets after optimization as well as during search to reduce the computation overhead related to hypervolume-based fitness assignment.

1 Introduction

In evolutionary multiobjective optimization, several challenges emerge when considering scenarios involving a large number of objectives – with respect to both search and decision making (Fleming et al. 2005; Hughes 2007; Sülflow et al. 2007). For tackling the resulting problems, some recent studies propose to automatically and adaptively reduce the number of objectives by means of corresponding dimensionality reduction techniques as suggested by Purshouse and Fleming (2003). One approach (Deb and Saxena 2006; Saxena and Deb 2007) uses principal component analysis (PCA) to determine a small subset of the objectives such that most of the variance of the solutions in the objective space is explained. Another approach (Brockhoff and Zitzler 2007b, 2009) was proposed by the authors of this paper and differs fundamentally from the work by Deb and Saxena (2006) and Saxena and Deb (2007) in the way the information loss caused by the dimensionality

D. Brockhoff (✉)
Computer Engineering and Networks Laboratory, ETH Zurich, 8092 Zurich, Switzerland
e-mail: dimo.brockhoff@tik.ee.ethz.ch

D. Jones et al. (eds.), *New Developments in Multiple Objective and Goal Programming*, Lecture Notes in Economics and Mathematical Systems 638,
DOI 10.1007/978-3-642-10354-4_6,

reduction is quantified; it aims at changing the underlying dominance structure as little as possible by minimizing the δ-error (Brockhoff and Zitzler 2007b, 2009), i.e., the maximum error that one makes in wrongly assuming that one solution dominates another one. Both approaches reduce the objective set by omitting selected objectives.

In this paper, we generalize the ideas of objective reduction in Brockhoff and Zitzler (2007b, 2009) by considering aggregations of several objectives and thereby make use of the fact that when aggregating objectives instead of omitting them less information is lost, i.e., the δ-error can be further decreased. In this context, the main goal can be restated as follows: find a minimum set of new objectives where each of them represents a weighted sum of the original objectives such that the dominance structure between solutions is (mostly) preserved. Clearly, this formulation also captures the omission of objectives as weights can be set to 0. We present a greedy algorithm to approximate the optimal solution to the problem of finding a minimum objective subset that preserves most of the dominance structure. It works by iteratively selecting a pair of objectives that is integrated into a new objective using weighted-sum aggregation. To validate the usefulness of the proposed approach, we apply it, on the one hand, to analyze and visualize high-dimensional Pareto set approximations and, on the other hand, to speed up the search process in hypervolume-based multiobjective optimization as suggested in Brockhoff and Zitzler (2007a). The experimental results indicate that especially with a large number of objectives the new method can better preserve the problem characteristics and has advantages over the existing methods.

2 Objective Reduction by Aggregating Objectives

In a multiobjective setting, k objective functions $f_i : X \to Z$ $(1 \leq i \leq k)$ that map a solution $\mathbf{x} \in X$ from the decision space X to its objective vector $f(\mathbf{x}) = (f_1(\mathbf{x}), \ldots, f_k(\mathbf{x}))$ have to be optimized simultaneously. We consider, without loss of generality, minimization problems only. The goal is to compute or approximate the set of Pareto optimal solutions with respect to a given dominance structure. Here, we assume that the underlying dominance structure is given by the weak Pareto dominance relation $\preceq_{\mathcal{F}'}$, i.e., a solution $\mathbf{x} \in X$ is at least as good as or *weakly dominating* a solution $\mathbf{y} \in X$ $(\mathbf{x} \preceq_{\mathcal{F}'} \mathbf{y})$ with respect to a set $\mathcal{F}'$ of objective functions if $\forall f_i \in \mathcal{F}' : f_i(\mathbf{x}) \leq f_i(\mathbf{y})$. We call two solutions *incomparable* if neither weakly dominates the other one, *comparable* if one dominates the other, and *indifferent* if they are mutually dominating each other. A solution $\mathbf{x}^* \in X$ is called *Pareto optimal* if for any other solution $\mathbf{x} \in X$, $\mathbf{x}^*$ is either weakly dominating $\mathbf{x}$ or is incomparable to $\mathbf{x}$ with respect to the set of all objectives. The set of all Pareto optimal solutions is called *Pareto(-optimal) set*, for which an approximation is sought.

Although the entire set of objectives $\mathcal{F} := \{f_1, \ldots, f_k\}$ is of interest, the reduction to a smaller set $\mathcal{F}' \subset \mathcal{F}$ might be necessary in practice, e.g., in terms of decision

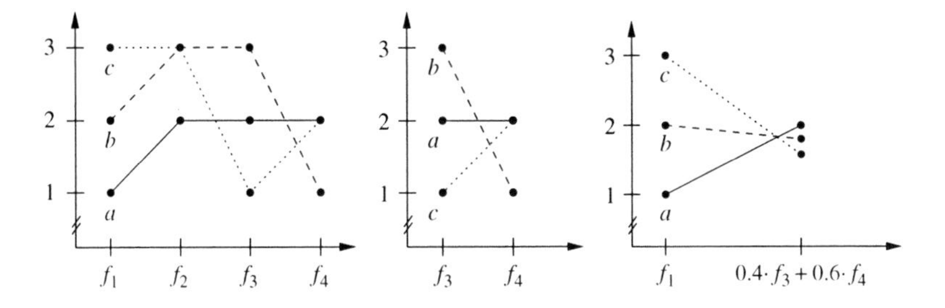

Fig. 1 Parallel coordinates plot for the three solutions **a** (*solid*), **b** (*dashed*), and **c** (*dotted*) in Examples 1–3: (*left*) original objectives; (*middle*) with respect to the objective subset $\{f_3, f_4\}$; (*right*) original objective f_1 and aggregated objective $0.4f_3 + 0.6f_4$. For details, see text

making where a small set of objectives can be easier taken into account than many objectives or in terms of search when the computation time highly depends on the number of objectives.[1] In this case of *objective reduction*, the question arises how the weak dominance relation changes if objectives are omitted. This question has been already answered in detail in previous studies (Brockhoff and Zitzler 2007b, 2009; Brockhoff et al. 2007): the only possible changes if objectives are omitted are that comparable solutions can become indifferent and incomparable solutions can become comparable or indifferent. In other words, only new comparabilities can be introduced by omitting objectives.

Example 1. *The left-hand side of Fig. 1 shows the parallel coordinates plot of three solutions* **a** *(solid line),* **b** *(dashed) and* **c** *(dotted) that are pairwisely incomparable. If the first objective f_1 is omitted, all three solutions remain mutually incomparable, i.e., the problem structure is preserved. If we, however, omit any two objectives, the dominance relation between the three solutions will change, e.g., if we omit in addition to f_1 also f_2, the solution* **a** *becomes weakly dominated by solution* **c** *considering minimization of the objectives, cf. the middle plot of Fig. 1.*

To quantify how much the dominance relation is changed, the δ-error has been proposed in Brockhoff and Zitzler (2007b, 2009), which will be explained in the following and used throughout the paper. If we change the set of objectives under consideration to the new set $\mathcal{F}'$, e.g., when omitting some objectives or aggregating them, we might wrongly assume for a solution pair that the first solution $\mathbf{x}$ is weakly dominating the second one $\mathbf{y}$ although this is not the case with respect to the entire set of objectives $\mathcal{F}$, cf. Example 1. In this case, we make an error for the solution pair $\mathbf{x}, \mathbf{y}$. The idea of the δ-error of Brockhoff and Zitzler (2007b, 2009) is the observation that we would not make an error by assuming $\mathbf{x} \preceq_{\mathcal{F}'} \mathbf{y}$ if the entire

[1] This is the case, e.g., in hypervolume-based evolutionary algorithms where the computation time per generation is exponential in the number of objectives (Brockhoff and Zitzler 2007a; Bringmann and Friedrich 2008). Also if the objective values have to be computed by time-consuming simulations, a lower number of objectives will speed up the evaluation time.

set of objectives would induce the same relation already. In particular, if we make the objective vector of $\mathbf{x}$ better by an additional term of $\max_{f_i \in \mathcal{F}}\{f_i(\mathbf{x}) - f_i(\mathbf{y})\}$, the resulting solution is weakly dominating $\mathbf{y}$ with respect to all objectives and the relation between the two solutions is preserved when changing to the objective set $\mathcal{F}'$. The resulting relation when we change the objective values as described above, is the additive ε-dominance relation of Laumanns et al. (2002). We say a solution $\mathbf{x}$ is weakly ε-dominating another solution $\mathbf{y}$ with respect to the entire objective set $\mathcal{F}$ if $\forall f_i \in \mathcal{F} : f(\mathbf{x}) - \varepsilon \leq f(\mathbf{y})$. With this additive ε-dominance relation, the δ-error for a solution pair can be defined as in Brockhoff and Zitzler (2007b, 2009) as the smallest δ value $\delta = \max_{f_i \in \mathcal{F}}\{f_i(\mathbf{x}) - f_i(\mathbf{y})\}$ such that $\mathbf{x}$ δ-dominates $\mathbf{y}$ with respect to the entire objective set $\mathcal{F}'$ if $\mathbf{x}$ weakly dominates $\mathbf{y}$ with respect to the new objective set $\mathcal{F}'$. The maximum of this error over all solution pairs is referred to as the *maximum δ-error* of the new objective set $\mathcal{F}'$ in Brockhoff and Zitzler (2007b, 2009). In addition, we propose the *average δ-error* here which is defined as the δ-error averaged over all solution pairs:

Definition 1. Given two sets of objectives $\mathcal{F}'$ and $\mathcal{F}$ and a set of solutions $A \subseteq X$, the maximum δ-error of Brockhoff and Zitzler (2007b, 2009) is defined as

$$\delta_{\max}(A, \mathcal{F}', \mathcal{F}) = \max_{\substack{\mathbf{x}, \mathbf{y} \in A \\ \mathbf{x} \not\preceq_{\mathcal{F}} \mathbf{y} \\ \mathbf{x} \preceq_{\mathcal{F}'} \mathbf{y}}} \max_{f_i \in \mathcal{F}}\{f_i(\mathbf{x}) - f_i(\mathbf{y})\}$$

and the average δ-error is defined as

$$\delta_{\text{avg}}(A, \mathcal{F}', \mathcal{F}) = \frac{1}{n(n-1)} \sum_{\substack{\mathbf{x}, \mathbf{y} \in A \\ \mathbf{x} \not\preceq_{\mathcal{F}} \mathbf{y} \\ \mathbf{x} \preceq_{\mathcal{F}'} \mathbf{y}}} \max_{f_i \in \mathcal{F}}\{f_i(\mathbf{x}) - f_i(\mathbf{y})\}.$$

Note that if both error measures are meant or if it is clear from the context which measure we are referring to, we will use the term δ-error in the remainder.

Example 2. *Consider again the solutions* **a** *and* **c** *of Fig. 1 if the objectives f_1 and f_2 are omitted. Since solution* **c** *is now weakly dominating* **a** *although it is not weakly dominating* **a** *with respect to all objectives we make an error of $\delta = \max_{f_i \in \mathcal{F}}\{f_i(\mathbf{c}) - f_i(\mathbf{a})\} = 2$. If we consider the maximum δ-error over all solution pairs the error is $\delta_{max} = 2$ for the set $\{f_3, f_4\}$ of remaining objectives. Note that no pair of objectives yields a smaller maximum error than $\delta_{max} = 1$ which is reached for the sets $\{f_1, f_3\}$ and $\{f_2, f_3\}$. In terms of the average δ-error, the set $\{f_3, f_4\}$ yields an error of $\delta_{avg} = 2/6$ – we only assume once a wrong dominance relation, namely for $\mathbf{c} \preceq_{\{f_3, f_4\}} \mathbf{a}$.*

We have seen that the omission of objectives introduces new comparabilities and information about the original dominance structure gets lost. As an alternative, the number of objectives can also be reduced by the aggregation of objectives and thereby keeping more of the information about the original objectives as we will see in the following. In this study, we consider the simplest case of objective aggregation: the weighted sum approach. An *aggregated objective* is a linear combination of the original ones $f_i^{\mathrm{a}} = \sum_{f_j \in \mathcal{F}} w_{i,j} f_j$ where we assume that the non-negative weights $w_{i,j}$ $(1 \leq j \leq k)$ for each of the new objectives sum up to 1. This formalism also contains the omission of objectives as an aggregation where all weights are either 1 or 0.

Example 3. *Once again, we consider the example of Fig. 1. As we have seen above, no objective pair can preserve the weak Pareto-dominance relation entirely. However, if we allow the aggregation of objectives, a set of two aggregated objectives can be found that preserves the dominance relation completely. The right-hand plot of Fig. 1 shows the parallel coordinate plot for such a set* $\{f_1, 0.4f_3 + 0.6f_4\}$. *All three solutions are still pairwisely incomparable, i.e., the original dominance relation is preserved.*

When aggregating objectives, the only change in the dominance structure is the introduction of comparabilities – exactly as with the omission of objectives. If for a pair $\mathbf{a}, \mathbf{b} \in X$ of solutions, $\mathbf{a} \preceq_{\mathcal{F}} \mathbf{b}$ holds, $\mathbf{b}$ will stay dominated with respect to any set of aggregated objectives. If, however, for a pair $\mathbf{a} \not\preceq_{\mathcal{F}} \mathbf{b}$ holds, the aggregation can introduce the domination of $\mathbf{a}$ if for all newly introduced objectives $f_i^{\mathrm{a}}(\mathbf{a}) \not> f_i^{\mathrm{a}}(\mathbf{b})$ holds due to the choice of the weights.

In Brockhoff and Zitzler (2007b, 2009), given a set $A \subseteq X$ of solutions and a $k_a \in \mathbb{N}$, the problem of finding the best objective subset $\mathcal{F}' \subseteq \mathcal{F}$ of size k_a that minimizes the resulting (maximum) δ-error was introduced as the k_a-EMOSS problem. Here, we generalize this problem to finding the best set of k_a aggregated objectives such that the resulting δ-error is minimized and denote it as the `OptimalAggregationProblem`:

Definition 2. Given a $k_a \in \mathbb{N}$, a set $A \subseteq X$ of solutions, and a chosen δ-error (either $\delta = \delta_{\max}$ or $\delta = \delta_{\mathrm{avg}}$). Let $\mathcal{F} = \{f_1, \ldots, f_k\}$ be the set of all objectives. The `OptimalAggregationProblem` with respect to δ is defined as follows: Find a set of weight vectors $W = \{\mathbf{w} = (w_1, \ldots, w_k) \in [0, 1]^k \mid \sum_{1 \leq i \leq k} w_i = 1\}$ with $|W| = k_a$ such that the δ-error of the set of aggregated objective vectors

$$\delta\Big(A, \bigcup_{(w_1,\ldots,w_k) \in W} \Big\{\sum_{i=1}^{k} w_i f_i\Big\}, \mathcal{F}\Big)$$

is minimal.

In the remainder of the paper, we will refer to the problem of finding the optimal aggregation with respect to the maximum error as $\mathtt{OA}_{\mathtt{max}}$ and denote the optimal aggregation problem with respect to the average δ-error as $\mathtt{OA}_{\mathtt{avg}}$. For both problems, we propose a greedy heuristic in the following.

3 A Greedy Heuristic for Finding the Best Aggregation

As the special case of k_a-EMOSS is already NP-hard (Brockhoff and Zitzler 2007b, 2009), we assume that the generalized problems $\texttt{OA}_{\texttt{max}}$ and $\texttt{OA}_{\texttt{avg}}$ are also too complex to solve them exactly. To solve the $\texttt{OA}_{\texttt{max}}$ and $\texttt{OA}_{\texttt{avg}}$ problems, we, therefore, propose a greedy approximation algorithm the idea of which is to iteratively aggregate objective pairs until the desired number of objectives is reached – in other words, the algorithm resembles the approach of hierarchical clustering.

Note that other non-greedy approximation algorithms for the problem of finding the best aggregation might be developed as well. For the special case of the k_a-EMOSS problem, López Jaimes et al. (2008) for example proposed to use feature selection algorithms known from machine learning recently. These algorithms have been shown to result in smaller δ-errors than the simple greedy algorithms proposed in Brockhoff and Zitzler (2007b, 2009). However, in accordance with Brockhoff and Zitzler (2007b, 2009), we propose only simple greedy algorithms here to show the principles and the usefulness of the approach of objective aggregation – while keeping in mind that advanced algorithms have to be developed, applied and tested in future work.

3.1 Main Procedure

Algorithm 1 shows the pseudo code of the aggregation procedure. Starting with the original objectives, i.e., with k weight vectors containing exactly one 1-entry and otherwise zeros (W), the δ-error equals 0. Then, in each step of the while-loop, the objective pair the aggregation of which yields the smallest error is aggregated and the corresponding weight vectors and the δ-error are adjusted. To this end, for each objective pair, represented by the weight vectors in W, the weight α when optimally aggregating this objective pair is computed. Optimally in this case, means that the δ-error is minimized when deleting the objective pair[2] $\mathbf{q}, \mathbf{r}$ and adding a new objective $f_{\text{new}} = \alpha\mathbf{q} + (1-\alpha)\mathbf{r}$. How the optimal weight α can be computed in the function $\texttt{aggregateOptimally}(A, \mathcal{F}', \mathcal{F})$ such that the δ-error is minimized will be explained in detail in the following.

3.2 Optimally Aggregating Two Objectives

Let us consider, without loss of generality, the case that f_1 and f_2 in a set of objectives $\mathcal{F}'$ have to be aggregated optimally, i.e., we have to find the weight $\alpha \in [0,1]$ such that the δ-error between the original objective set $\mathcal{F}$ and the new set $\mathcal{F}' \setminus \{f_1, f_2\} \cup f_{\text{new}}$ is minimal where $f_{\text{new}} = \alpha f_1 + (1-\alpha) f_2$ is the new aggregated

---

[2] For simplicity, we use the term objective and the corresponding weight vector interchangeably throughout the paper.

Algorithm 1 A greedy heuristic for the problems $\texttt{OA}_{\texttt{max}}$ and $\texttt{OA}_{\texttt{avg}}$

Require: solution set $A \subseteq X$ with set of objectives $\mathcal{F} = \{f_1, \ldots, f_k\}$
 number of desired objectives $k_a \geq 1$
 Init:
 $W = \{\mathbf{w} \in \mathbb{R}^k \mid \sum_{i=1}^k w_i = 1 \wedge \exists i \in \{1, \ldots, k\} : w_i = 1\}$
 $\delta = 0$
 while $|W| > k_a$ **do**
  $\delta_{\text{best}} = +\infty$
  for all $\mathbf{q}, \mathbf{r} \in W, \mathbf{q} \neq \mathbf{r}$ **do**
   $\alpha = \texttt{aggregateOptimally}(\mathbf{q}, \mathbf{r}, A, W, \mathcal{F})$
   $W' = W \setminus \{\mathbf{q}, \mathbf{r}\} \cup \{\alpha\mathbf{q} + (1-\alpha)\mathbf{r}\}$
   $\delta' = \delta_{\text{max/avg}}\left(A, \bigcup_{(w'_1, \ldots, w'_k \in W')} \{\sum_{i=1}^k w'_i f_i\}, \mathcal{F}\right)$
   if $\delta' \leq \delta_{\text{best}}$ **then**
    $W_{\text{best}} = W'$
    $\delta_{\text{best}} = \delta'$
   end if
  end for
  $W = W_{\text{best}}$
  $\delta = \delta_{\text{best}}$
 end while
 return (W, δ)

objective. In the following, we will use for the current set of objectives excluding the two objectives that have to be aggregated the term *remaining objectives* and call the objective set $\mathcal{F}'_{\text{rem}} := \mathcal{F}' \setminus \{f_1, f_2\}$.

The idea behind the function $\texttt{aggregateOptimally}(A, \mathcal{F}'_{\text{rem}} \cup \{f_{\text{new}}\}, \mathcal{F})$ is to determine for each solution pair $\mathbf{a}, \mathbf{b} \in A$ a function $\Delta_{(\mathbf{a},\mathbf{b})} : [0,1] \to \mathbb{R}_0^+$ that gives for each possible weight $\alpha \in [0,1]$ the δ-error that is introduced if the objective pair is aggregated to the new objective f_{new}. Figure 3 gives some examples how this $\Delta_{(\mathbf{a},\mathbf{b})}$ function can look like. How this function $\Delta_{(\mathbf{a},\mathbf{b})}$ can be computed is explained later on in detail. The best weight α_{opt} over all solution pairs and the corresponding δ-error δ_{opt} can then be computed as

$$\alpha_{\text{opt}} = \underset{\alpha \in [0,1]}{\operatorname{argmin}} \Delta_A(\alpha),$$
$$\delta_{\text{opt}} = \min_{\alpha \in [0,1]} \Delta_A(\alpha),$$

where, depending on the problem to solve,

$$\Delta_A(\alpha) = \Delta_{A,\max}(\alpha) = \max_{\mathbf{a},\mathbf{b} \in A} \Delta_{(\mathbf{a},\mathbf{b})}(\alpha)$$

for $\texttt{OA}_{\texttt{max}}$ and

$$\Delta_A(\alpha) = \Delta_{A,\text{avg}}(\alpha) = \frac{1}{|A|(|A|-1)} \sum_{\mathbf{a},\mathbf{b} \in A} \Delta_{(\mathbf{a},\mathbf{b})}(\alpha)$$

for $\texttt{OA}_{\texttt{avg}}$. In other words, if the error depending on the weight α is known for all solution pairs, the maximum δ-error $\Delta_{A,\max}$ is computed as the maximum over all solution pairs whereas the average δ-error $\Delta_{A,\text{avg}}$ is the δ-error averaged over all solution pairs. The optimal weight is then chosen in the best weight interval, i.e., the weight interval with the minimal δ-error, see again Fig. 3 for an illustration. We would like to mention already here, that the choice of α_{opt} is not unique – most of the time it is rather an *optimal weight interval* than a single value. We will discuss the actual choice of α_{opt} in the interval with smallest δ-error later and decide to fix the center of the optimal interval as the optimal weight for the moment.

Now, we explain how to determine the function $\Delta_{(\mathbf{a},\mathbf{b})}$. To this end, we fix $\mathbf{a}$ and $\mathbf{b}$ and distinguish between two situations: (a) $\delta(\{\mathbf{a},\mathbf{b}\},\mathcal{F}'_{\text{rem}},\mathcal{F}) = 0$, i.e., even if we omit the objectives f_1 and f_2, we make no error. In this case, α can be chosen arbitrarily in $[0,1]$ and the δ-error is 0, i.e., $\forall \alpha \in [0,1] : \Delta_{(\mathbf{a},\mathbf{b})}(\alpha) = 0$. (b) $\mathbf{a}$ and $\mathbf{b}$ are standing with respect to the remaining objectives $\mathcal{F}'_{\text{rem}}$ in a different relationship than with respect to the entire objective set $\mathcal{F}$, i.e., it depends on the choice of α which error we make. In this case (i) $\mathbf{a} \preceq_{\mathcal{F}'_{\text{rem}}} \mathbf{b}$ but $\mathbf{a} \not\preceq_{\mathcal{F}} \mathbf{b}$ and/or (ii) $\mathbf{b} \preceq_{\mathcal{F}'_{\text{rem}}} \mathbf{a}$ but $\mathbf{b} \not\preceq_{\mathcal{F}} \mathbf{a}$ can hold.

Assume first that (i) holds but not (ii). In this case, we need to choose α such that $\mathbf{a} \not\preceq_{\mathcal{F}'} \mathbf{b}$ holds with respect to the new objective set $\mathcal{F}' = \mathcal{F}'_{\text{rem}} \cup \{f_{\text{new}}\}$ to make no error, or in other words, we need to ensure that $f_{\text{new}}(\mathbf{a}) > f_{\text{new}}(\mathbf{b})$. This inequality can be rewritten as

$$
\begin{aligned}
& f_{\text{new}}(\mathbf{a}) > f_{\text{new}}(\mathbf{b}) \\
& \iff \alpha f_1(\mathbf{a}) + (1-\alpha) f_2(\mathbf{a}) > \alpha f_1(\mathbf{b}) + (1-\alpha) f_2(\mathbf{b}) \\
& \iff \alpha(f_1(\mathbf{a}) - f_1(\mathbf{b}) + f_2(\mathbf{b}) - f_2(\mathbf{a})) > f_2(\mathbf{b}) - f_2(\mathbf{a}) \qquad (1)
\end{aligned}
$$

yielding – depending on the precise objective values of $\mathbf{a}$ and $\mathbf{b}$ – an interval $S \subseteq [0,1]$ of α, where the δ-error is zero. For all other choices of α, we make an error that is the same as if we would omit f_1 and f_2 entirely. Therefore

$$
\Delta^{(i)}_{(\mathbf{a},\mathbf{b})} = \begin{cases} 0 & \text{if } \alpha \in S, \\ \delta_{\text{max/avg}}(\{\mathbf{a},\mathbf{b}\},\mathcal{F}'_{\text{rem}},\mathcal{F}) & \text{else,} \end{cases}
$$

where the δ error in the else case again depends on the problem to solve. Examples of this function for different solution pairs can be seen in Fig. 3.

The case where (ii) but not (i) holds follows analogously by changing the roles of $\mathbf{a}$ and $\mathbf{b}$ and yields a similar function $\Delta^{(ii)}_{(\mathbf{a},\mathbf{b})}$ as above.

What remains to investigate is the case where both (i) and (ii) hold. In this case, we make an error with any choice of α: either $\mathbf{a} \preceq_{\{f_{\text{new}}\}} \mathbf{b}$ and the error is $\Delta^{(i)}_{(\mathbf{a},\mathbf{b})}$ as in case (i) or $\mathbf{b} \preceq_{\{f_{\text{new}}\}} \mathbf{a}$ and the error is $\Delta^{(ii)}_{(\mathbf{a},\mathbf{b})}$. In case both $\mathbf{a} \preceq_{\{f_{\text{new}}\}} \mathbf{b}$ and $\mathbf{b} \preceq_{\{f_{\text{new}}\}} \mathbf{a}$ holds, i.e., if $\mathbf{a}$ and $\mathbf{b}$ are indifferent with respect to f_{new}, the resulting error is the maximum of both errors $\Delta^{(i)}_{(\mathbf{a},\mathbf{b})}$ and $\Delta^{(ii)}_{(\mathbf{a},\mathbf{b})}$. Thus,

$$
\Delta_{(\mathbf{a},\mathbf{b})}(\alpha) = \max\{\Delta_{(\mathbf{a},\mathbf{b})}\} \quad .
$$

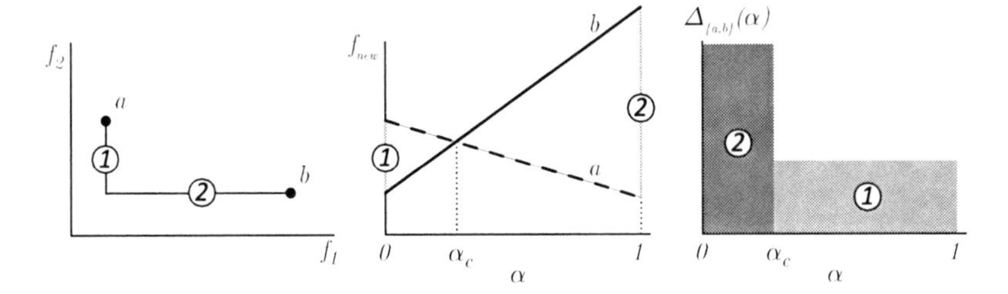

Fig. 2 Illustration of aggregation of two objectives that is optimal for two solutions a and b: (*left*) original objectives; (*middle*) resulting aggregated objective values $f_{\text{new}} = \alpha f_1 + (1-\alpha) f_2$ for all choices of $0 \le \alpha \le 1$; (*right*) corresponding $\delta_{\max}$-error for all choices of α

Example 4. *The left-hand plot of Fig. 2 shows two solutions* **a** *and* **b** *with two objectives f_1, f_2, the aggregation of which to the new objective $f_{new} = \alpha f_1 + (1-\alpha) f_2$ causes an error for all choices of $\alpha \in [0,1]$. The reason why we make an error for all choices of α is that the two solutions become comparable for any choice of α although they are incomparable with respect to the original objectives. The resulting dominance relation changes from* $\mathbf{b} \preceq_{\{f_{new}\}} \mathbf{a}$ *to* $\mathbf{a} \preceq_{\{f_{new}\}} \mathbf{b}$ *when α is increased, cf. the middle plot of Fig. 2. The value α_c for which both solutions are indifferent (the intersection point α_c in the middle plot of Fig. 2), can be computed when the inequality in (1) is changed to an equality. The resulting δ-errors for each choice of α (indicated as "①" and "②" in Fig. 2) are shown in the rightmost plot of Fig. 2.*

Note that in the biobjective Example 4, the set of remaining objectives is empty. In the following example, the set of three objectives has to be reduced by one objective as it would be the case within a run of Algorithm 1, i.e., the set of remaining objectives always contains one objective. In addition, we consider more than one solution pair here.

Example 5. *Consider four solutions* $\mathbf{a}, \mathbf{b}, \mathbf{c}, \mathbf{d} \in A$ *with the objective vectors* $f(\mathbf{a}) = (1,8,4)$, $f(\mathbf{b}) = (6,2,7)$, $f(\mathbf{c}) = (3,4,4)$, *and* $f(\mathbf{d}) = (0,7,7)$. *In the following, we illustrate the progress of Algorithm 1 if the average δ-error has to be minimized and the objective set has to be reduced to $k_a = 2$ objectives. After initialization, the algorithm computes the optimal choice of α for each objective pair. Figure 3 shows the computation for the objective pair (f_1, f_2) in more detail: for each solution pair* $\mathbf{x}, \mathbf{y}$, *the* $\Delta_{(\mathbf{x},\mathbf{y})}$ *function is computed (left plot of Fig. 3) and the average over all these functions is taken as the overall error function Δ_A (right-hand plot of Fig. 3). The optimal choice of α has to lie in the interval indicated by the arrow in the right-hand plot of Fig. 3. For the other two objective pairs (f_1, f_3) and (f_2, f_3), similar Δ_A functions can be obtained, see Fig. 4. The best intervals with an error of $\delta_{avg} = 0$ can be obtained when aggregating f_1 and f_2 with an $\alpha \in (0.5, 0.65)$ and when aggregating f_2 and f_3 with an $\alpha \in (0.6, 0.65)$.*

Note that it is not specified in Algorithm 1 how the α value has to be chosen within an optimal interval. Although each choice yields the same δ-error in the current aggregation step, the choice within the optimal interval might influence the

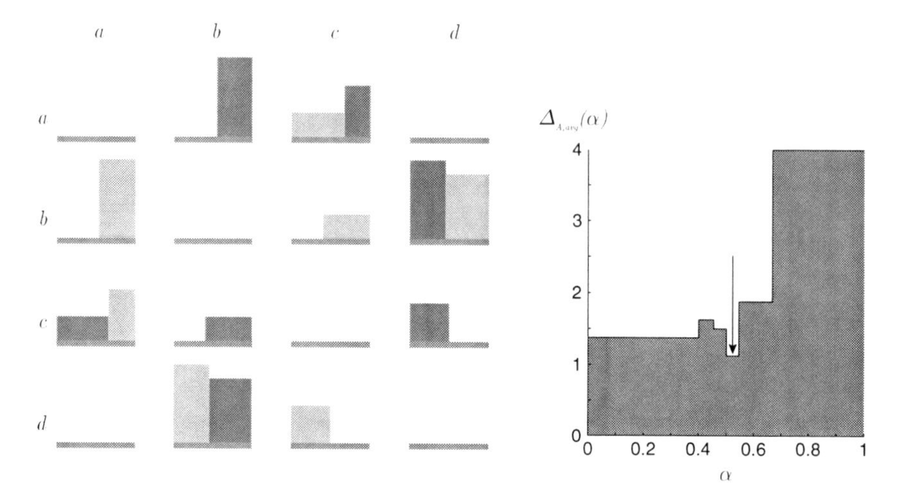

Fig. 3 Computation of δ-errors dependent on the weight for all solution pairs (*left*) and corresponding δ-error averaged over all solution pairs (*right*) if the first two objectives in Example 5 are aggregated. The *arrow* in the right-hand plot indicates the optimal weight interval

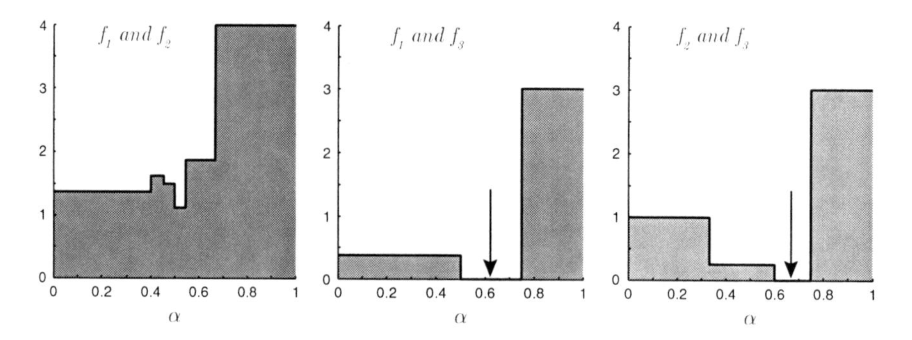

Fig. 4 Illustration of the δ-errors averaged over all solutions for the objective pairs f_1, f_2 (*left*), f_1, f_3 (*middle*), and f_2, f_3 (*right*) of Example 5. The optimal intervals are indicated by *arrows*

errors we make when aggregating other objectives in a future step of the algorithm. Therefore, we compare different strategies experimentally in Sect. 4.1.

4 Experimental Validation

After presenting the aggregation heuristic for the $\texttt{OA}_{\texttt{max}}$ and $\texttt{OA}_{\texttt{avg}}$ problems, three main questions remain open: (1) is it important how to choose the weight within the optimal interval found by the aggregation heuristics? (2) how much can the δ-error be reduced by the new aggregation heuristics in comparison to simply omitting objectives? and (3) what can be gained by aggregating objectives instead of omitting them during the search? This section is addressing these three questions experimentally before we apply the proposed aggregation heuristics afterwards to a real-world problem of optimizing a radar waveform to show their usefulness with respect to visualization.

4.1 The Influence of Different Weight Choices Within the Optimal Interval

As discussed above, the weight choice within the optimal interval found by the proposed aggregation heuristics might have an influence on the overall δ-error although in the current step, all choices lead to the same δ-error. To investigate the influence of different weight choices on the overall resulting error, we compare four variants of the greedy $\texttt{OA}_{\texttt{max}}$ heuristic: If the optimal objective pair and the corresponding optimal weight interval is found, we either choose the weight in the middle of this interval (variant "CENTER"), uniformly at random within the interval ("UNIFORM"), or as the center of the left ("LEFT") or right ("RIGHT") half of the interval.

Settings: To perform the comparison, we created 3×51 random instances by choosing the objective values uniformly at random in $[0, 1]$: 51 solution sets of 50 solutions with 4 objectives, 51 sets of 100 solutions with 6 objectives, and 51 sets of 200 solutions with 8 objectives. These instances have been handed over to the greedy $\texttt{OA}_{\texttt{max}}$ algorithm that had to reduce the objective set to 30% and 60% of the original objectives.[3] The resulting maximum δ-errors have been ranked for all four variants "CENTER", "UNIFORM", "LEFT", and "RIGHT" (rank 1: best, rank 4: worst) and then compared by means of the Kruskal–Wallis test with the additional Conover–Inman procedure for multiple comparisons as described in Conover (1999, pp. 288–290). All tests have been performed on the basis of a p-value of 0.05.

Results: Without any exception, no significant difference between the algorithm variants could be observed. Apparently, the greedy property of the algorithm, i.e., not looking ahead to weight choices in future steps of the algorithm, does not allow for different behavior between the four variants. In other words, a significantly better choice in one step might be outweighed by a worse choice in one of the next steps and vice versa. Since the four variants do not show significant differences, we decided to use the "CENTER" variant exclusively in the following due to its slightly faster implementation.

[3] More precisely, to 1 and 2 objectives for the 4-objective instances, to 1 and 3 for the 6-objective instances, and to 2 and 4 for the 8-objective instances.

4.2 Comparison Between Aggregation and Omission

To experimentally validate the proposed aggregation heuristics for the $\mathtt{OA}_{\mathtt{max}}$ and $\mathtt{OA}_{\mathtt{avg}}$ problems, we compare them with the k_a-EMOSS heuristic from Brockhoff and Zitzler (2007b, 2009) that, starting with an empty objective set, iteratively chooses the objective that minimizes the overall δ-error until the desired number of objectives is reached.

Settings: For the comparison, Pareto set approximations of 18 test problem instances have been generated by running the Indicator-Based Evolutionary Algorithm IBEA, proposed in Zitzler and Künzli (2004), for 100 generations with the standard settings of the PISA package (Bleuler et al. 2003). The only parameter that changed over the different problem instances was the population size which was chosen as 100 for the 5-objective problems and as 200 for the 15-objective problems. In addition to the DTLZ2, DTLZ3, DTLZ7 (Deb et al. 2005),[4] WFG3, WFG6, WFG8 (Huband et al. 2006)[5] and three instances of the 0-1-knapsack problem with 100 (KP100), 250 (KP250), and 500 (KP500) items (Laumanns et al. 2004) with 5 and 15 objectives each, we also considered two instances of a network processor design problem called EXPO (Künzli et al. 2004) with 3 and 4 objectives. The Pareto set approximations for the EXPO instances had 43 solutions and 143 solutions respectively.

Based on these test instances, the three considered greedy algorithms for the k_a-EMOSS, the $\mathtt{OA}_{\mathtt{max}}$, and the $\mathtt{OA}_{\mathtt{avg}}$ problem respectively were used to reduce the objectives to 90%, 60%, and 30% of the number of original objectives, i.e., to 1, 3, and 4 for the 5-objective problems and to 4, 9, and 13 for the 15-objective problems. The EXPO instances had to be reduced to 1 and 2 (3-objective) and to $1, 2$, and 3 objectives (4-objective) respectively. Table 1 shows the resulting normalized[6] δ-errors for all three algorithms and, in addition, the δ-error averaged over all solution pairs for the new heuristics proposed in Sect. 3.

Results: For all considered problem instances, except when the 15-objective DTLZ2 problem is reduced to 4 objectives ($\lfloor 15 \cdot 30\% \rfloor = 4$), the aggregation heuristic optimizing the maximum error yields lower or the same errors than the heuristic that omits objectives. Therefore, we conclude that the error can, in general, be decreased for the same number of objectives when aggregation is allowed. For example, the preservation of the entire dominance relation ($\delta = 0$) can be achieved for the reduction to 60% of the original objectives for 4 of the 12 DTLZ and knapsack problem instances if aggregation is allowed whereas the k_a-EMOSS heuristic cannot find objective subsets of this sizes without making an error. Note that the only case

[4] The number of decision variables has been set to 250.

[5] For all WFG problems, the number of decision variables has been also fixed to 250 and the number of position variables has been chosen to 168 and the number of distance variables to 82 such that it can be kept constant over all numbers of objectives.

[6] The δ-errors have been normalized to the objective values of each instance such that the difference between the highest and lowest objective value equals 1 for every objective.

Table 1 Comparison of the aggregation heuristics with the greedy k_a-EMOSS heuristic of Brockhoff and Zitzler (2007b, 2009). The table entries show for the greedy k_a-EMOSS algorithm the maximum δ-errors whereas for the aggregation heuristics both the maximum (left of slash) and the average (right of slash) δ-errors are shown

	# Objectives	# Solutions	Greedy k_a-EMOSS			Greedy aggregation maximum δ-error			Greedy aggregation average δ-error		
			30%	60%	90%	30%	60%	90%	30%	60%	90%
KP100	5	100	0.9263	0.5164	0.4860	0.7829/0.2659	0.3781/0.0072	0.2576/0.0003	1.0000/0.2649	0.4911/0.0060	0.2576/0.0003
KP100	15	200	0.8180	0.3483	0.0000	0.7705/0.0209	0.0000/0.0000	0.0000/0.0000	0.8977/0.0130	0.0000/0.0000	0.0000/0.0000
KP250	5	100	0.8588	0.6967	0.2797	0.8588/0.3149	0.2758/0.0013	0.1189/0.0001	0.8156/0.3049	0.6300/0.0014	0.1189/0.0000
KP250	15	200	0.7622	0.3421	0.0000	0.6921/0.0200	0.0000/0.0000	0.0000/0.0000	0.7807/0.0090	0.0000/0.0000	0.0000/0.0000
KP500	5	100	0.7484	0.5041	0.2370	0.6829/0.1924	0.3638/0.0117	0.2079/0.0028	0.6735/0.1934	0.3773/0.0074	0.2370/0.0018
KP500	15	200	0.6425	0.4350	0.2775	0.4333/0.0297	0.1991/0.0012	0.0866/0.0000	0.5691/0.0175	0.2608/0.0008	0.1800/0.0000
DTLZ2	5	100	0.9909	0.9699	0.9202	0.9909/0.5876	0.7928/0.0113	0.6371/0.0019	1.0000/0.5592	0.8618/0.0091	0.6541/0.0010
DTLZ2	15	200	0.9418	0.8910	0.0000	0.9517/0.0524	0.4044/–[a]	0.0000/0.0000	0.9823/0.0119	0.4044/–[a]	0.0000/0.0000
DTLZ5	5	100	0.9523	0.9062	0.8958	0.9368/0.5324	0.6323/0.0077	0.4771/0.0010	0.9794/0.5171	0.8017/0.0077	0.4897/0.0007
DTLZ5	15	200	0.8601	0.8030	0.0000	0.8549/0.0226	0.0000/0.0000	0.0000/0.0000	0.9286/0.0083	0.0000/0.0000	0.0000/0.0000
DTLZ7	5	100	0.1353	0.1335	0.1321	0.1353/0.1121	0.1222/0.0003	0.0000/0.0000	0.1558/0.1116	0.1233/0.0002	0.0000/0.0000
DTLZ7	15	200	0.0778	0.0700	0.0000	0.0748/0.0001	0.0000/0.0000	0.0000/0.0000	0.0778/0.0001	0.0000/0.0000	0.0000/0.0000
WFG3	5	100	0.6611	0.0000	0.0000	0.6611/0.2402	0.0000/0.0000	0.0000/0.0000	0.6611/0.2402	0.0000/0.0000	0.0000/0.0000
WFG3	15	200	0.0000	0.0000	0.0000	0.0000/0.0000	0.0000/0.0000	0.0000/0.0000	0.0000/0.0000	0.0000/0.0000	0.0000/0.0000
WFG6	5	100	0.1329	0.0000	0.0000	0.1329/0.0490	0.0000/0.0000	0.0000/0.0000	0.1329/0.0490	0.0000/0.0000	0.0000/0.0000
WFG6	15	200	0.0000	0.0000	0.0000	0.0000/0.0000	0.0000/0.0000	0.0000/0.0000	0.0000/0.0000	0.0000/0.0000	0.0000/0.0000
WFG8	5	100	0.5636	0.0000	0.0000	0.5636/0.1783	0.0000/0.0000	0.0000/0.0000	0.5636/0.1741	0.0000/0.0000	0.0000/0.0000
WFG8	15	200	0.0000	0.0000	0.0000	0.0000/0.0000	0.0000/0.0000	0.0000/0.0000	0.0000/0.0000	0.0000/0.0000	0.0000/0.0000
EXPO	3	43	0.9423	0.9423	0.6901	0.9423/0.3010	0.9423/0.3010	0.3729/0.0104	1.0000/0.2537	1.0000/0.2537	0.7754/0.0096
EXPO	4	143	0.6665	0.5049	0.2202	0.6665/0.1684	0.3248/0.0213	0.1197/0.0017	0.8048/0.1414	0.5484/0.0050	0.1281/–[a]

[a] The error is here smaller than 0.00005, where errors of 0.0000 are exactly zero

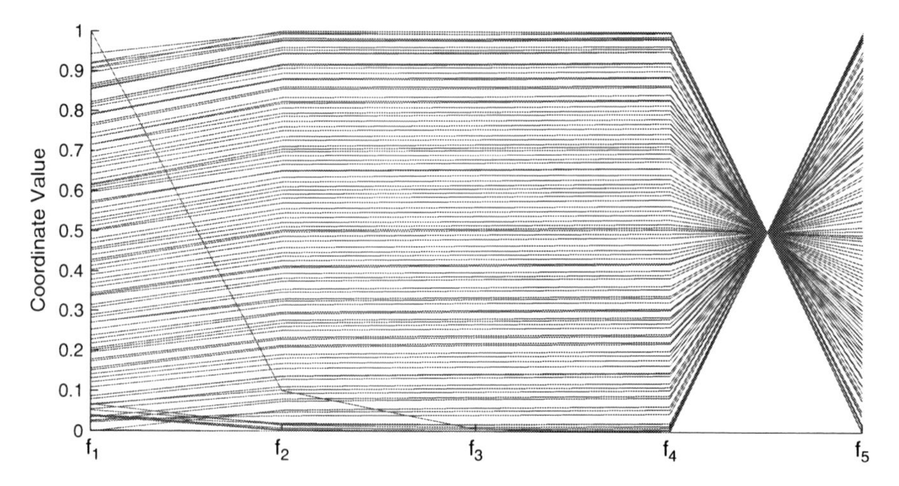

Fig. 5 Normalized parallel coordinates plot of the used WFG3 instance with 5 objectives and 100 solutions

where the omission of objectives yields a smaller error than both aggregation heuristics occurs for a reduction to 30%, i.e., a small number of objectives. This might be due to the fact that the greedy heuristic for objective omission creates the objective subset by *adding* objectives greedily instead of reducing the number of objectives step-by-step as the greedy aggregation heuristics do.

As expected, the aggregation heuristic optimizing the average error performs better with respect to the average error and worse with respect to the maximum error in most cases compared to the heuristic optimizing the maximum error. The next section will show that a low average δ-error is beneficial when visualizing solution sets of many-objective problems.

Furthermore, we would like to point out that the Pareto set approximations of the WFG test problem instances are seemingly of a certain shape where only a few objectives are necessary to describe the trade-offs between the objectives. Figure 5 shows as an example the used set of objective vectors for the WFG3 instance with 5 objectives in a normalized parallel coordinates plot. By looking at the objectives f_4 and f_5, we can observe that this objective pair induces all incomparabilities between the solutions due to their opposing nature in the found region of the search space. Therefore, the results of the WFG test problems in Table 1 might not be as representative for real-world applications as for the other problems.

4.3 Objective Reduction During Search

After showing how the proposed objective reduction algorithms can be applied in decision making scenarios, it remains to investigate whether aggregating the objectives automatically can also be beneficial during search. Purshouse and Fleming

(2003) have been the first to propose the idea of using objective reduction techniques during search – mainly to deal with problems such as the lower selection pressure or the more complicated density estimation if many objectives have to be optimized simultaneously. Especially if the computation time of the optimizer highly depends on the number of objectives, e.g., in hypervolume-based search algorithms where the exact calculation of the hypervolume indicator is exponential in the number of objectives,[7] objective reduction seems to be highly beneficial. Similar to the study in Brockhoff and Zitzler (2007a), where omitting objectives during search has been shown to significantly improve hypervolume-based algorithms, we investigate in the following what can be gained by using the proposed aggregation heuristic within a hypervolume-based evolutionary algorithm instead of simply omitting objectives.

Basic Algorithm The basis of this study is the Simple Indicator-Based Evolutionary Algorithm (SIBEA) proposed in Zitzler et al. (2007) which uses the hypervolume indicator I_H, originally proposed in Zitzler and Thiele (1998), to guide the search.[8] The algorithm works as follows. After the initial population P is formed by μ randomly selected solutions, new generations are performed until a given time limit T is reached. A generation starts with a random selection of μ solutions in P that are then recombined and mutated to μ offsprings. These offsprings are inserted in the population P and the population of the next generation is determined by the following procedure: after a non-dominated sorting of P, the non-dominated fronts are, starting with the best front, completely inserted into the new population P' until the size of P' is at least μ. For the first front F the inclusion of which yields a population size larger than μ, the solutions $\mathbf{x}$ in this front with the smallest indicator loss $d(\mathbf{x}) := I_H(F) - I_H(F \setminus \{x\})$ are successively removed from the new population where the indicator loss is recalculated every time a solution is removed.

In addition, SIBEA can apply various objective reduction strategies to improve the running time of the hypervolume computation. To this end, every G generations an objective reduction is performed, i.e., it is decided which objectives are chosen for optimization and which ones are neglected during the next G generations. This objective reduction has already been shown to be beneficial during search due to the high computation time for the hypervolume losses in high dimensions (Brockhoff and Zitzler 2007a). The incorporation of the greedy k_a-EMOSS algorithm of Brockhoff and Zitzler (2007b, 2009) and the aggregation heuristics for the OA_{max} and OA_{avg} problems yields three modified versions of SIBEA. Every $G = 50$ generations, we compute the best objective subset (for the k_a-EMOSS based $\text{SIBEA}_{k_a\text{-EMOSS}}$) or the best aggregation (versions denoted by $\text{SIBEA}_{\text{max}}$ and $\text{SIBEA}_{\text{avg}}$) on the current population and consider only the computed objectives

[7] The #$\mathcal{P}$-hardness proof of Bringmann and Friedrich (2008) implies that no exact polynomial algorithm for the hypervolume indicator exists unless $\mathcal{P} = \mathcal{NP}$.

[8] The hypervolume indicator or $\mathcal{S}$-measure of a solution set $A \subset X$ is informally defined as the space that is dominated by the solutions in A which itself is dominating a reference set. Here, we use only a single reference point and refer to Beume et al. (2007) for an exact definition. The hypervolume indicator has always to be maximized.

for the next 50 generations of SIBEA. We use the greedy k_a-EMOSS algorithm based SIBEA$_{k_a\text{-EMOSS}}$ version for the comparison with the new aggregation approach since it showed the best performance in the exhaustive comparison of Brockhoff and Zitzler (2007a).

Settings The comparison of the three versions is performed with the following settings. 11 runs for each combination of problem [scaled DTLZ2$_{BZ}$, scaled DTLZ3$_{BZ}$, and DTLZ7 as in Brockhoff and Zitzler (2007a) with 5, 10, and 15 objectives] and objective set size ($k_a = 2, 3, 4$) are performed where the computation time is set to $T = 15$ min on a 64-bit AMD Linux machine with four cores. Afterwards, the hypervolume indicator values of the populations after the time T has been reached are computed with respect to all objectives and the non-parametric Kruskal–Wallis test followed by the Conover–Inman procedure for multiple testing (see pages 288–290 of Conover 1999) is used to support the hypothesis that one algorithm "systematically" produces larger hypervolume indicator values than another one by ranking all values and comparing the rank sums. The significance level has been set to $p = 0.05$ and the reference points of the hypervolume computation are chosen as $r_{\text{DTLZ2}} = (50, \ldots, 50)$, $r_{\text{DTLZ3}} = (25000, \ldots, 25000)$, and $r_{\text{DTLZ7}} = (170, \ldots, 170)$. Table 2 shows both the ranking of the mean values (in brackets) and a ranking given by the outcomes of the statistical tests: for each algorithm A, the number of other algorithms that statistically outperform A is shown. For both rankings, lower numbers are better. Figure 6 shows, in addition, the box plots of the achieved hypervolume values for the problem instances with 15 objectives where the statistical tests do not support the hypothesis of differences in the mean values.

To also compare the runs with different numbers of aggregated objectives against each other, we decided to run the algorithms again for 11 runs with different random seeds and compare all algorithms for all objective subset sizes against each other for each of the three problems and each number of original objectives. The same statistical Conover–Inman test after the mandatory Kruskal–Wallis test has been performed; test problem instances and reference points are the same as before. Table 3 shows the rankings of the mean of the hypervolume indicator values (again in brackets) and the number of algorithms that produce significantly higher hypervolume values as before – now by comparing all 9 different algorithms for each number of original objectives.

Results The results of the statistical tests support the main conclusion that the aggregation has some advantages over the omission of objectives especially if the objective set is reduced to only two objectives. However, the advantage diminishes when more objectives are involved during the search. Except for the DTLZ3$_{BZ}$ problem with 10 objectives, the omission heuristic always performs better with respect to the mean values than the aggregation heuristics if the objective set is reduced to 4 objectives. In addition, we can observe from the box plots in Fig. 6 that the omission heuristic becomes better with increasing k_a whereas both aggregation heuristics become better when the size of the reduced objective set is decreased. One explanation for that is the high running time of the aggregation heuristics: the running time of 15 min is mainly used for deriving the aggregation in every 50th generation.

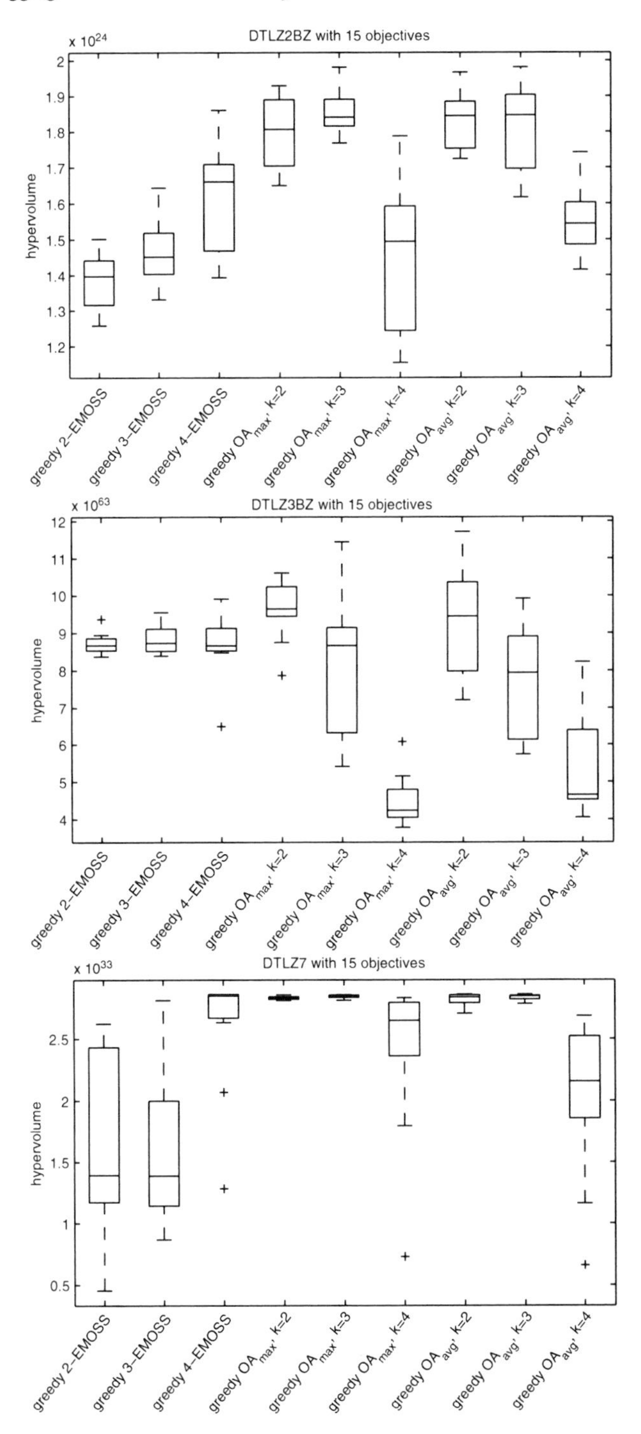

Fig. 6 Boxplots of hypervolume indicator values for the objective reduction heuristics applied during search: (*top*) $DTLZ2_{BZ}$; (*middle*) $DTLZ3_{BZ}$; (*bottom*) DTLZ7 with 15 objectives each

Table 2 Ranking of the hypervolume indicator values for the SIBEA versions with greedy k_a-EMOSS heuristic ($I_{H,k_a\text{-EMOSS}}$) and the aggregation heuristics with maximum ($I_{H,\max}$) and average δ-error ($I_{H,\text{avg}}$) based on the Kruskal–Wallis test with subsequent Conover–Inman procedure when the three algorithms are compared for each pair of original objective number and the number of aggregated objectives. The rank corresponds to the number of algorithms that significantly produce better hypervolume indicator values. In addition, the ranking of the means is given in brackets. Lower values are always better

# Original objectives	# Aggregated objectives	Scaled DTLZ2BZ			Scaled DTLZ3BZ			DTLZ7		
		$I_{H,k_a\text{-EMOSS}}$	$I_{H,\max}$	$I_{H,\text{avg}}$	$I_{H,k_a\text{-EMOSS}}$	$I_{H,\max}$	$I_{H,\text{avg}}$	$I_{H,k_a\text{-EMOSS}}$	$I_{H,\max}$	$I_{H,\text{avg}}$
5	2	2(3)	0(2)	0(1)	2(3)	0(2)	0(1)	2(3)	0(2)	0(1)
5	3	0(1)	1(2)	1(3)	0(1)	2(3)	0(2)	0(1)	2(3)	1(2)
5	4	0(1)	2(3)	1(2)	0(1)	1(3)	1(2)	0(1)	2(3)	1(2)
10	2	2(3)	1(2)	0(1)	2(3)	0(2)	0(1)	2(3)	0(2)	0(1)
10	3	2(3)	0(1)	0(2)	2(3)	0(1)	1(2)	2(3)	1(2)	0(1)
10	4	0(1)	1(2)	2(3)	1(2)	2(3)	0(1)	0(1)	1(2)	2(3)
15	2	2(3)	0(2)	0(1)	1(3)	0(1)	0(2)	2(3)	0(2)	0(1)
15	3	2(3)	0(2)	0(1)	0(1)	0(2)	0(3)	2(3)	0(1)	0(2)
15	4	0(1)	0(3)	0(2)	0(1)	2(3)	1(2)	0(1)	0(2)	2(3)

Table 3 Ranking of the hypervolume indicator values for the SIBEA versions with greedy k_a-EMOSS heuristic ($I_{H,k_a\text{-EMOSS}}$) and the aggregation heuristics with maximum ($I_{H,\max}$) and average δ-error ($I_{H,\text{avg}}$) over all number of aggregated objectives for each problem instance based on the Kruskal–Wallis test with subsequent Conover–Inman procedure. The rank corresponds to the number of algorithms that significantly produce better hypervolume indicator values and the ranking of the means are given in brackets

# Original objectives	# Aggregated objectives	Scaled DTLZ2BZ			Scaled DTLZ3BZ			DTLZ7		
		$I_{H,k_a\text{-EMOSS}}$	$I_{H,\max}$	$I_{H,\text{avg}}$	$I_{H,k_a\text{-EMOSS}}$	$I_{H,\max}$	$I_{H,\text{avg}}$	$I_{H,k_a\text{-EMOSS}}$	$I_{H,\max}$	$I_{H,\text{avg}}$
5	2	3(4)	0(1)	0(2)	7(9)	0(3)	0(1)	5(7)	0(3)	0(4)
5	3	0(3)	4(5)	4(6)	3(6)	1(4)	0(2)	0(1)	5(6)	2(5)
5	4	5(7)	8(9)	7(8)	3(5)	5(8)	3(7)	0(2)	7(9)	7(8)
10	2	8(9)	1(4)	0(2)	5(8)	0(3)	0(2)	6(8)	1(3)	0(1)
10	3	6(8)	0(1)	1(3)	5(7)	0(1)	2(4)	4(5)	1(4)	0(2)
10	4	4(5)	4(6)	5(7)	4(6)	6(9)	4(5)	4(6)	6(7)	8(9)
15	2	7(9)	0(3)	1(4)	1(6)	0(1)	0(2)	5(8)	0(4)	0(3)
15	3	7(8)	0(1)	0(2)	0(4)	1(5)	1(7)	2(7)	0(2)	0(1)
15	4	4(6)	4(5)	4(7)	0(3)	7(9)	7(8)	1(5)	4(6)	7(9)

For example on the DTLZ3BZ problem with 15 objectives, most of the aggregation runs are performing 200–350 generations only whereas almost all SIBEA$_{k_a\text{-EMOSS}}$ runs (except the ones running with four objectives) are able to run for 1,000 generations or more.

When comparing the algorithms over all possible numbers of desired objectives, it turns out that, except for the $DTLZ2_{BZ}$ with 15 objectives, no other algorithm produces significantly higher hypervolume values than $SIBEA_{avg}$ that reduces the number of objectives to $k_a = 2$. However, with respect to the mean hypervolume indicator values, this aggregation heuristic is assigned only twice the best rank and four times the second best. Nevertheless, we can also conclude in this comparison that the aggregation heuristics are often performing better and overall not worse than the greedy omission heuristic.

5 Application to a Real-World Problem

To show the advantages of the proposed aggregation heuristics with respect to visualization, we apply the objective omission approach of Brockhoff and Zitzler (2007b, 2009) and the aggregation heuristics to the problem of finding good waveforms for aircraft radars as proposed by Hughes (2007). The problem is formulated with 9 objectives in total and the set of more than 22,000 known non-dominated solutions builds the basis of our analysis. We follow the approach of Brockhoff and Zitzler (2007b, 2009) and reduce this set to 107 solutions that yields an ε-approximation of the entire set for $\varepsilon = 0.062$.

Visualizing the set of all known non-dominated solutions is a crucial task in decision making. For the radar waveform problem, the high number of both solutions and objectives makes the visualization difficult. Even the visualization as a parallel coordinates plot does not provide much information to the decision maker, see Hughes (2007) for the plot. Here, we argue that our objective reduction techniques can help to gain a detailed understanding of the problem itself by plotting lower dimensional projections of the non-dominated solutions. General dimensionality reduction techniques such as PCA do not take into account the Pareto-dominance relation between the solutions when reducing the dimensionality of data. Thus, the dominance relation is not preserved and many solutions dominate each other as can be seen in Table 4. In contrast, our approach of objective aggregation takes the Pareto-dominance relation into account and reduces the number of objectives while the dominance relation is changed as little as possible. Figure 7 shows the 2D plots of all known non-dominated solutions for the radar problem if different kinds of reduction techniques are applied.

The number of solution pairs that remain non-dominated increases from the PCA plot over the one with original objectives only to the aggregated objective plots of the maximum error and average error versions of the proposed greedy heuristic. Note that only PCA was applied directly to the set of all non-dominated solutions; for the other three reduction approaches, the reduced set of 107 points has been used – whereas the evaluation has been performed with respect to all points.

Table 4 Comparison of reduction to 2 objectives for the radar waveform problem with the reduced set of 107 points (*above*) and the set of all known non-dominated points (*below*): number of solution pairs (total 5,671 and $\approx 2.6 \times 10^8$ respectively) for that we wrongly assume comparability together with the maximum and average δ-errors

		# Comparable solution pairs	$\delta_{\max}$	δ_{avg}
107 solutions	Exact algorithm for k_a-EMOSS (Brockhoff and Zitzler, 2007b, 2009)	2,835	0.9935	0.111005
	PCA	2,525	0.9797	0.088250
	Greedy aggregation with maximum δ-error	872	0.8898	0.022609
	Greedy aggregation with average δ-error	585	0.9638	0.015443
all solutions	Exact algorithm for k_a-EMOSS	130,084,321	1.0000	0.087803
	PCA	134,147,513	0.9999	0.081457
	Greedy aggregation with maximum δ-error	49,189,796	0.9902	0.019601
	Greedy aggregation with average δ-error	32,471,066	0.9983	0.012444

6 Conclusions

Within this study, we proposed a generalization of a recently proposed framework for reducing the number of objectives in multiobjective problems with a high number of objectives. Instead of simply omitting the objectives, we allow to aggregate some of the objectives to achieve a reduced set of objectives that preserves most of the problem structure. The proposed heuristics to find a good aggregation have been compared to the previously presented approaches of objective omission and showed better performance in terms of objective set sizes during decision making and in terms of the achieved quality of the Pareto set approximations during optimization. Another aspect of many-objective problems, the visualization of high dimensional objective vectors, has also been covered in this paper by applying the proposed methods to a radar waveform problem.

Although we showed the usefulness of automatically finding a good aggregation of the original objectives both in decision making and search, the proposed aggregation heuristics might also be helpful in other scenarios, e.g., if an initial weighting of the objectives within the well-known weighted sum method is sought. Furthermore, the objective reduction algorithms can, in principle, be applied within any multiobjective evolutionary algorithm. However, the application of the proposed methods to these scenarios remains future work.

Acknowledgements The authors would like to thank Johannes Bader for the support with the illustrations. Dimo Brockhoff has been supported by the Swiss National Science Foundation (SNF) under grant 112079.

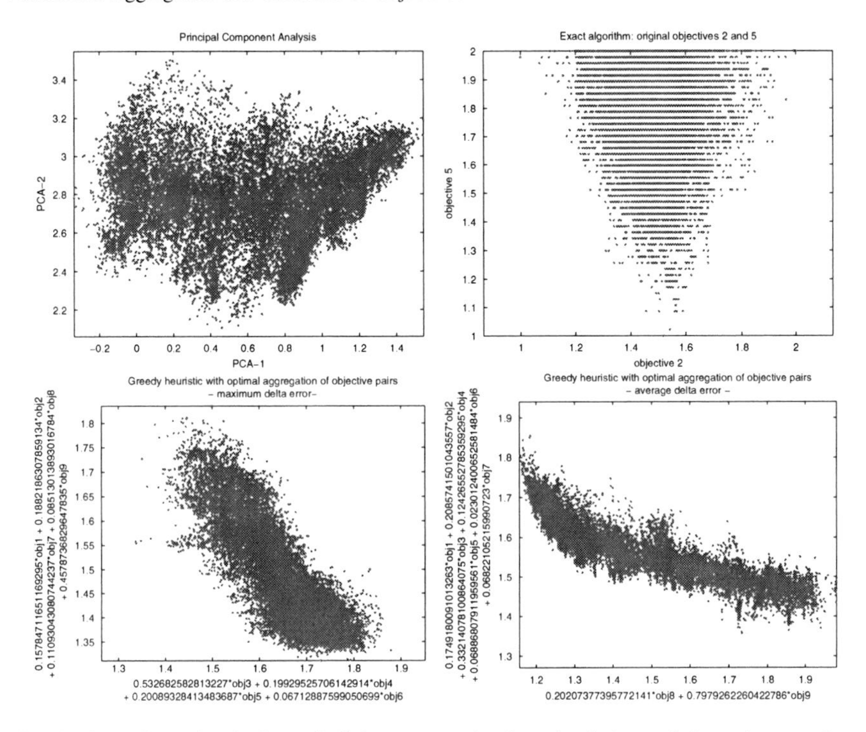

Fig. 7 Two-dimensional plots of all known non-dominated solutions of the radar waveform problem where the objectives are chosen with respect to PCA (*upper left*), the exact k_a-EMOSS algorithm from Brockhoff and Zitzler (2007b, 2009) (*upper right*), and the greedy aggregation algorithm with maximum error (*lower left*) and average error (*lower right*)

References

Beume N, Naujoks B, Emmerich M (2007) SMS-EMOA: multiobjective selection based on dominated hypervolume. Eur J Oper Res 181:1653–1669

Bleuler S, Laumanns M, Thiele L, Zitzler E (2003) PISA – a platform and programming language independent interface for search algorithms. In: Fonseca CM, et al (eds) Conference on evolutionary multi-criterion optimization (EMO 2003). LNCS, vol 2632. Springer, Berlin, pp 494–508

Bringmann K, Friedrich T (2008) Approximating the volume of unions and intersections of high-dimensional geometric objects. In: Hong SH, Nagamochi H, Fukunaga T (eds) International symposium on algorithms and computation (ISAAC 2008). LNCS, vol 5369. Springer, Berlin, pp 436–447

Brockhoff D, Zitzler E (2007a) Improving hypervolume-based multiobjective evolutionary algorithms by using objective reduction methods. In: Congress on evolutionary computation (CEC 2007). IEEE Press, pp 2086–2093

Brockhoff D, Zitzler E (2007b) Offline and online objective reduction in evolutionary multiobjective optimization based on objective conflicts. TIK Report 269, Computer Engineering and Networks Laboratory (TIK), ETH Zurich

Brockhoff D, Zitzler E (2009) Objective reduction in evolutionary multiobjective optimization: theory and applications. Evol Comput 17(2):135–166

Brockhoff D, Friedrich T, Hebbinghaus N, Klein C, Neumann F, Zitzler E (2007) Do additional objectives make a problem harder? In: Thierens D, et al (eds) Genetic and evolutionary computation conference (GECCO 2007). ACM Press, New York, NY, USA, pp 765–772

Conover WJ (1999) Practical nonparametric statistics, 3rd edn. Wiley, New York

Deb K, Saxena D (2006) Searching for Pareto-optimal solutions through dimensionality reduction for certain large-dimensional multi-objective optimization problems. In: Congress on evolutionary computation (CEC 2006). IEEE Press, pp 3352–3360

Deb K, Thiele L, Laumanns M, Zitzler E (2005) Scalable test problems for evolutionary multiobjective optimization. In: Abraham A, Jain R, Goldberg R (eds) Evolutionary multiobjective optimization: theoretical advances and applications, chap 6. Springer, Berlin, pp 105–145

Fleming PJ, Purshouse RC, Lygoe RJ (2005) Many-objective optimization: an engineering design perspective. In: Coello Coello CA, et al (eds) Conference on evolutionary multi-criterion optimization (EMO 2005). LNCS, vol 3410. Springer, Berlin, pp 14–32

Huband S, Hingston P, Barone L, While L (2006) A review of multiobjective test problems and a scalable test problem toolkit. IEEE Trans Evol Comput 10(5):477–506

Hughes EJ (2007) Radar waveform optimization as a many-objective application benchmark. In: Conference on evolutionary multi-criterion optimization (EMO 2007). LNCS, vol 4403. Springer, Berlin, pp 700–714

Künzli S, Bleuler S, Thiele L, Zitzler E (2004) A computer engineering benchmark application for multiobjective optimizers. In: Coello Coello CA, Lamont G (eds) Applications of multi-objective evolutionary algorithms. World Scientific, Singapore, pp 269–294

Laumanns M, Thiele L, Deb K, Zitzler E (2002) Combining convergence and diversity in evolutionary multiobjective optimization. Evol Comput 10(3):263–282

Laumanns M, Thiele L, Zitzler E (2004) Running time analysis of evolutionary algorithms on a simplified multiobjective Knapsack problem. Nat Comput 3(1):37–51

López Jaimes A, Coello Coello CA, Chakraborty D (2008) Objective reduction using a feature selection technique. In: Conference on genetic and evolutionary computation (GECCO 2008). ACM, New York, USA, pp 673–680, doi http://doi.acm.org/10.1145/1389095.1389228

Purshouse RC, Fleming PJ (2003) Conflict, harmony, and independence: relationships in evolutionary multi-criterion optimisation. In: Conference on evolutionary multi-criterion optimization (EMO 2003). LNCS, vol 2632. Springer, Berlin, pp 16–30

Saxena DK, Deb K (2007) Non-linear dimensionality reduction procedures for certain large-dimensional multi-objective optimization problems: employing correntropy and a novel maximum variance unfolding. In: Conference on evolutionary multi-criterion optimization (EMO 2007). LNCS, vol 4403. Springer, Berlin, pp 772–787, doi 10.1007/978-3-540-70928-2_58

Sülflow A, Drechsler N, Drechsler R (2007) Robust multi-objective optimization in high dimensional spaces. In: Conference on evolutionary multi-criterion optimization (EMO 2007). LNCS, vol 4403. Springer, Berlin, pp 715–726, doi 10.1007/978-3-540-70928-2_54

Zitzler E, Künzli S (2004) Indicator-based selection in multiobjective search. In: Yao X, et al (eds) Conference on parallel problem solving from nature (PPSN VIII). LNCS, vol 3242. Springer, Berlin, pp 832–842

Zitzler E, Thiele L (1998) Multiobjective optimization using evolutionary algorithms – a comparative case study. In: Conference on parallel problem solving from nature (PPSN V), Amsterdam, pp 292–301

Zitzler E, Brockhoff D, Thiele L (2007) The hypervolume indicator revisited: on the design of Pareto-compliant indicators via weighted integration. In: Obayashi S, et al (eds) Conference on evolutionary multi-criterion optimization (EMO 2007). LNCS, vol 4403. Springer, Berlin, pp 862–876

Trade-Off Analysis in Discrete Decision Making Problems Under Risk

Maciej Nowak

Abstract The paper considers a discrete stochastic multi-attribute decision making problem. This problem is defined by a finite set of alternatives A, a set of attributes X and a set E of evaluations of alternatives with respect to the attributes. In the stochastic case the evaluation of each alternative with respect to each attribute is characterized by a random variable. Thus, the comparison of two alternatives leads to the comparison of two vectors of probability distributions. In the paper a new interactive procedure for solving this problem is proposed. At each iteration a candidate alternative is proposed to the decision maker. If he/she is satisfied with the proposal, the procedure ends. Otherwise, the decision maker is asked to select the attribute to be improved and the attributes that can be decreased, ordered lexicographically starting with the one to be decreased first. The relations between distributions of trade-offs are used to generate a new proposal. An example is presented to illustrate the proposed technique.

1 Introduction

Interactive approach is probably the most often used method for solving multi-attribute decision making problems. It assumes that the decision maker (DM) is capable of defining attributes that influence his/her preferences and to provide preference information with respect to a given solution or a given set of solutions (local preference information). Two main advantages are usually mentioned for employing interactive techniques. First, such methods need much less a priori information about the DM's preferences. Second, as the DM is closely involved in all phases of the problem solving process, he/she puts much reliance in the generated solution, and as a result, the final solution has a better chance of being implemented.

M. Nowak
Department of Operations Research, The Karol Adamiecki University of Economics in Katowice, ul. 1 Maja 50, 40–287 Katowice, Poland
e-mail: Maciej.Nowak@ae.katowice.pl

D. Jones et al. (eds.), *New Developments in Multiple Objective and Goal Programming*, Lecture Notes in Economics and Mathematical Systems 638,
DOI 10.1007/978-3-642-10354-4_7,

The kind of local preference information required varies for each interactive procedure. Two main paradigms are employed when the information about the DM's preferences is collected: direct and indirect. According to the first one, the DM expresses his/her preferences in relation to the values of attributes. Such approach is used in techniques proposed by Benayoun et al. (1971), Wierzbicki (1980), Steuer (1986), and Spronk (1981). Indirect collection of preferences means that the decision maker has to determine the trade-offs among attributes at each iteration, given the current candidate solution. The classical method by Geoffrion et al. (1972) is an example of such approach. These two classes are not disjoint. The methods proposed by Zionts and Wallenius (1976) and Kaliszewski and Michalowski (1999) combine both approaches.

This paper focuses on discrete multi-attribute decision making problems under risk. By "discrete" we mean that a finite number of alternatives are explicitly known. Evaluations of the alternatives with respect to attributes are characterized by random variables. Various approaches have been proposed for such problem. Keeney and Raiffa (1976) suggest multiattribute utility function approach for this problem. They show that if the additive independence condition is verified, then a multiattribute comparison of two alternatives can be decomposed into one-attribute comparisons. In practice, however, both the estimation of one-attribute utility functions and the assessment of the synthesis function are difficult. Saaty and Vargas (1987) proposed a version of the AHP that introduces uncertainty. Various techniques based on the outranking approach were also suggested: Dendrou et al. (1980), Martel et al. (1986), and D'Avignon and Vincke (1988).

In this paper stochastic dominance (SD) rules are used for comparing distributional evaluations. Huang et al. (1978) showed that if the additive independence condition is verified, then the necessary condition for multi-attribute stochastic dominance (MSD) is the verification of stochastic dominance with respect to each attribute. In practice the MSD rule is very rarely verified. Zaras and Martel (1994) suggested weakening the unanimity condition and accepting a majority attribute condition. They proposed MSD_r – multiattribute stochastic dominance for a reduced number of attributes. This approach is based on the observation that people tend to simplify the multiattribute problem by taking into account only the most important attributes. The procedure consists of two steps. First, the SD relations are verified for each pair of alternatives with respect to all attributes. Next, the multiattribute aggregation is realized – the ELECTRE I methodology is used to obtain the final ranking of alternatives.

Interactive procedures for discrete multi-attribute decision making problems based on stochastic dominance have been proposed in Nowak (2004, 2006). The first is an extension of the STEM method. In each step a candidate alternative, which has a minimal distance to the ideal solution, is generated. A min–max rule is used for measuring this distance. The decision maker examines the evaluations of the candidate alternative with respect to attributes and selects the one that satisfies him/her. Then the limit of concessions, which can be made on average evaluations with respect to this attribute, is defined. The procedure continues until a satisfactory solution is found. The INSDECM procedure proposed in Nowak (2006) combines the SD approach and mean-risk analysis. It is assumed that the decision maker is able to express his/her requirements defining the restrictions based both on average evaluations and on scalar risk measures.

Sometimes the DM is not able to express his/her preferences directly – by defining the minimum or maximum values for one or more distribution characteristics. Often he/she is able to choose only that attribute which should be improved and the attributes that can be decreased without defining the limits of such concessions. In such a case trade-offs can be used for identifying a new proposal. The aim of this paper is to propose an interactive technique using dialog scenario of this type.

The paper is structured as follows. The problem is formulated in Sect. 2. Section 3 provides basic information about point-to-point trade-offs. In Sect. 4 an interactive procedure is presented. The next section gives a numerical example. The last section consists of conclusions.

2 Formulation of the Problem

The decision situation considered in this paper may be conceived as a problem (**A**, **X**, **E**) where **A** is a finite set of alternatives $a_i, i = 1, 2, \ldots, m$, **X** is a finite set of attributes $X^p, p = 1, 2, \ldots, n$, and **E** is a set of evaluations of projects with respect to attributes $X_i^p, i = 1, 2, \ldots, m, p = 1, 2, \ldots, n$. We assume that the attributes are defined in such a way that a larger value is preferred to a smaller one.

This work focuses on decision making problems under risk. Thus, we will assume that the evaluation of a_i with respect to X^p is a random variable with a cumulative probability distribution function $F_i^p(x)$ defined as follows:

$$F_i^p(x) = \Pr\left(X_i^p \leq x\right)$$

The attributes are supposed to be probabilistically independent, and are also supposed to satisfy the preference independence condition. Thus, the overall comparison of two alternatives can be decomposed into one-attribute comparisons of probability distributions.

Two main approaches are usually used for such comparisons: mean-risk models and stochastic dominance. The former is based on two criteria: one measuring expected outcome and another one representing variability of outcomes. The latter is based on an axiomatic model of risk-averse preferences and leads to conclusions that are consistent with the axioms. In fact, mean-risk approaches are not capable of modeling even the entire gamut of risk-averse preferences. Moreover, for typical statistics used as risk measures, the mean-risk approach may lead to inferior conclusions (Ogryczak and Ruszczyński 1999).

In this paper we will use stochastic dominance rules for modeling preferences of the DM in relation to each attribute. First stochastic dominance and second stochastic dominance are defined as follows:

Definition 1. (FSD – First Degree Stochastic Dominance).
X_i^p dominates X_j^p by FSD rule ($X_i^p \succ_{\text{FSD}} X_j^p$) if and only if
$F_i^p(x) \neq F_j^p(x)$ and $F_i^p(x)$-$F_j^p(x) \leq 0$ for all $x \in \mathbf{R}$

Definition 2. (SSD – Second Degree Stochastic Dominance).
X_i^p dominates X_j^p by FSD rule ($X_i^p \succ_{\text{SSD}} X_j^p$) if and only if

$$F_i^p(x) \neq F_j^p(x) \text{ and } \int_{-\infty}^{x} \left(F_i^p(y)\text{-}\ F_j^p(y)\right) dy \leq 0 \text{ for all } x \in \mathbf{R}$$

The FSD is the most general relation. If $X_i^p \succ_{\text{FSD}} X_j^p$, then X_i^p is preferred to X_j^p within all models preferring larger outcomes. The use of SSD requires more restrictive assumptions. If $X_i^p \succ_{\text{SSD}} X_j^p$, then X_i^p is preferred to X_j^p within all risk-averse preference models that prefer larger outcomes.

In this paper we assume that the DM is risk averse. Thus we will assume that X_i^p dominates X_j^p by stochastic dominance rule ($X_i^p \succ_{\text{SD}} X_j^p$) if $X_i^p \succ_{\text{FSD}} X_j^p$ or $X_i^p \succ_{\text{SSD}} X_j^p$. We will use this rule for comparing evaluations of alternatives with respect to attributes, and for analyzing relations between distributions of point-to-point trade-offs.

3 Point-to-Point Trade-Offs

A trade-off is defined for a particular solution and for a selected pair of the attributes. It specifies the amount by which the value of one attribute increases while that of the other one decreases when a particular solution is replaced by another given solution.

Let us start with a decision making problem under certainty. For a pair of alternatives a_i and a_j and a pair of attributes X^p and X^q, a point-to-point trade-off is the ratio of a relative value increase in one attribute (X^p) per unit of value decrease in the reference attribute (X^q) when the alternative a_i is replaced by the alternative a_j.

$$T_{ji}^{pq} = \frac{X_j^p - X_i^p}{X_i^q - X_j^q}$$

Let us assume that the DM analyzes the alternative a_i and decides that the evaluation with respect to X^p should be improved, while the evaluation with respect to X^q can be decreased. In this case we will look for alternatives a_j such that $X_j^p > X_i^p$ and $X_j^q \geq X_i^q$, and choose the one for which the increase of X^p is maximal. If such alternatives do not exist, then an alternative maximizing point-to-point trade-off will be proposed.

In the stochastic case the situation is much more complicated, as various situations have to be taken into account when a point-to-point trade-off for a pair of alternatives (a_j, a_i) and a pair of attributes (X^p, X^q) is computed. In fact, such a trade-off is a random variable whose distribution is a mixture of four distributions:

$X_i^p, X_i^q, X_j^p, X_j^q$. In this paper we will assume that the attributes are probabilistically independent and satisfy independence conditions allowing us to use an additive utility function. To generate probability distribution of point-to-point trade-off T_{ji}^{pq} we have to analyze the following cases:

1. $X_j^p > X_i^p$ and $X_i^q > X_j^q$,
2. $X_j^p > X_i^p$ and $X_i^q = X_j^q$,
3. $X_j^p > X_i^p$ and $X_i^q < X_j^q$,
4. $X_j^p = X_i^p$ and $X_i^q > X_j^q$,
5. $X_j^p = X_i^p$ and $X_i^q = X_j^q$,
6. $X_j^p = X_i^p$ and $X_i^q < X_j^q$,
7. $X_j^p < X_i^p$ and $X_i^q > X_j^q$,
8. $X_j^p < X_i^p$ and $X_i^q = X_j^q$,
9. $X_j^p < X_i^p$ and $X_i^q < X_j^q$.

Only the first case describes the classical trade-off situation. Cases (2) and (3) describe situations in which it is possible to improve the value of X^p without decreasing X^q. For such situations we will assume that $T_{ji}^{pq} = M$, where M is a "big number". If (4), (5), or (6) takes place, then we will assume that $T_{ji}^{pq} = 0$, as replacing a_i by a_j will not change the value of X^p. And finally for cases (7), (8), and (9) we will assume that $T_{ji}^{pq} = -M$, as replacing a_i by a_j will decrease the value of X^p. Thus for given $X_i^p, X_i^q, X_j^p, X_j^q$, the value of trade-off will be computed as follows:

$$T_{ji}^{pq} = \begin{cases} \frac{X_j^p - X_i^p}{X_i^q - X_j^q} & \text{if } X_j^p > X_i^p \text{ and } X_i^q > X_j^q \\ M & \text{if } X_j^p > X_i^p \text{ and } X_i^q \le X_j^q \\ 0 & \text{if } X_j^p = X_i^p \\ -M & \text{if } X_j^p < X_i^p \end{cases}$$

Let us consider the following example. Alternatives a_1 and a_2 are evaluated with respect to attributes X^1 and X^2. Distributions of alternatives with respect to attributes are presented in Table 1. To generate the distribution of T_{21}^{12}, we have to consider all possible combinations of the values $X_1^1, X_1^2, X_2^1, X_2^2$ (Table 2).

Table 1 Example 1 – evaluations of alternatives

Distributions for X^1			Distributions for X^2		
	a_1	a_2		a_1	a_2
100			20		0.5
150	0.5	0.25	30	0.75	
200	0.5		40		0.5
250		0.75	50	0.25	

Table 2 Example 1 – generation of distribution of T_{21}^{12}

	X_1^1	X_2^1	X_1^2	X_2^2	Prob.	T_{21}^{12}
1	100	150	30	20	0.046875	5.00
2	100	150	30	40	0.046875	M
3	100	150	50	20	0.015625	1.67
4	100	150	50	40	0.015625	5.00
5	100	250	30	20	0.140625	15.00
6	100	250	30	40	0.140625	M
7	100	250	50	20	0.046875	5.00
8	100	250	50	40	0.046875	15.00
9	200	150	30	20	0.046875	–M
10	200	150	30	40	0.046875	–M
11	200	150	50	20	0.015625	–M
12	200	150	50	40	0.015625	–M
13	200	250	30	20	0.140625	5.00
14	200	250	30	40	0.140625	M
15	200	250	50	20	0.046875	1.67
16	200	250	50	40	0.046875	5.00

Let us again assume that the DM analyzes the alternative a_i and decides that the evaluation with respect to X^p should be improved, while the evaluation with respect to X^q can be decreased. Assuming that a set of potential new proposals has been generated, the following question arises: how can distributions of point-to-point trade-offs be compared to identify a new proposal? In this paper SD rules are employed for comparison of these distributions. We will assume that the decision-maker is risk-averse, and as a result, FSD and SSD rules can be used for the analysis of relations between distributions of point-to-point trade-offs.

4 The Procedure

The main ideas of the procedure are as follows:

- A candidate for most preferred solution is presented to the DM at each iteration
- If the DM is satisfied with the proposal – the procedure ends
- Otherwise – the DM is asked to select the attribute to be improved and the attributes that can be decreased, ordered lexicographically starting with the one to be decreased first
- Information about relations between trade-offs distributions is used to generate a new candidate

To start the procedure we have to identify the first proposal. In the approach presented here, SD rules and min–max criterion are employed in this phase. The first proposal is identified in the following steps:

1. Identify SD relations between distributional evaluations for each pair of alternatives and for each attribute.
2. For each alternative compute:

$$\overline{d}_i = \max_{p \in \{1,\ldots,n\}} \{d_i^p\}$$

where:

$$d_i^p = \text{card} \quad D_i^p$$
$$D_i^p = \{a_j : X_j^p \succ_{\text{SD}} \; X_i^p\}$$

3. Choose the alternative for which $\bar{d}_i$ is minimal.

In our problem the evaluations of alternatives with respect to attributes are expressed by probability distributions. In such a case it is not easy for the DM to compare alternatives. On the one hand, the DM is usually interested in maximizing the expected outcomes, on the other hand, however, he/she finds the variability of outcomes very important as well. In the approach presented here, as in the INSDECM procedure (Nowak 2006), it is assumed that the decision maker is able to specify the method of data presentation. For each attribute he or she may choose one or more scalar measures to be presented to him or her. Both expected outcome measures (mean, median, mode) and variability measures (standard deviation, semideviation, probability of getting outcomes not greater or not less than target value) can be chosen. Moreover, the DM may change his/her mind while the procedure is in progress, and specify other sets of measures at successive iterations. For example, while initially the DM may be interested mainly in the expected outcomes, in subsequent phases of the procedure he/she may focus on risk measures.

Let us denote:

$\mathbf{A}^{(l)}$ – the set of alternatives considered at iteration l, $\mathbf{A}^{(1)} = \mathbf{A}$
$\mathbf{B}$ – the set of potential new proposals
a_s – the candidate alternative

At each iteration the following steps are executed:

1. Ask the DM to specify the data he/she is interested in – the parameters of distributional evaluations such as mean, standard deviation, probability of getting a value not less (not greater) than ξ, etc.
2. Compute values of parameters for each alternative under consideration, identify the best value of each parameter.
3. Present the data to the DM:
   - The values of parameters for the candidate alternative a_s
   - Best values of parameters attainable within the set of alternatives
4. Ask the DM whether he/she is satisfied with the proposal. If the answer is YES – the procedure ends – the proposal is assumed to be the final solution of the problem.

5. If the DM is not satisfied with the proposal, ask him/her to specify the attribute be improved first and to set the order of the remaining attributes, starting from the one that can be decreased first. Let p be the number of the attribute that the DM would like to improve, while $\{q_1, q_2, \ldots, q_{n-1}\}$ is the order of the attributes that can be decreased.
6. Identify the set of alternatives satisfying the requirements expressed by the DM:

$$\mathbf{A}^{(l+1)} = \left\{a_i : a_i \in \mathbf{A}^{(l)}, a_i \neq a_s, \neg X_s^p \succ_{\text{SD}} X_i^p\right\}$$

If the set $\mathbf{A}^{(l+1)}$ is empty, notify the DM that it is not possible to find an alternative satisfying his/her requirements, unless previous restrictions are relaxed. Then ask the DM whether he/she would like to relax the previous requirements. If the answer is NO, return to 5. Otherwise, generate the set of alternatives to be considered in the next phases of the procedure:

$$\mathbf{A}^{(l+1)} = \left\{a_i : a_i \in \mathbf{A}^{(1)}, a_i \neq a_s, \neg X_s^p \succ_{\text{SD}} X_i^p\right\}$$

7. Assume: $\mathbf{B} = \mathbf{A}^{(l+1)}$, $k = 1$.
8. Generate probability distributions of trade-offs $T_{i\,s}^{p\,q_k}$ for each i such that $a_i \in \mathbf{B}$.
9. Compare distributions of trade-offs with respect to SD rules and identify the set of non-dominated distributions. If the number of non-dominated distributions is equal to 1, assume the corresponding alternative to be the new proposal and go to 13.
10. Identify the alternatives with dominated trade-offs and exclude them from the set **B**.
11. If $k < n - 1$, assume $k := k + 1$ and go to 8.
12. The trade-offs for each pair of attributes have been compared, and the set of potential new proposals **B** still consists of more than one alternative. As the analysis of trade-offs has not provided a clear recommendation for the new proposal, analyze the relations between alternatives with respect to attributes. Start from attribute X^p and identify the set of alternatives with non-dominated evaluations according to SD rules. If the number of such alternatives is equal to 1, assume the corresponding alternative to be a new proposal and go to 13. Otherwise exclude from **B** the alternatives that are dominated according to SD rules with respect to attribute X^p. Next, analyze relations with respect to other attributes. In this phase of the procedure use a reversed lexicographic order of attributes: $q_{n-1}, q_{n-2}, \ldots, q_1$. For each attribute identify the dominated alternatives using SD rules and exclude them from **B**. Continue until **B** consists of one alternative. If all attributes have been considered and **B** still consists of more than one alternative, assume any of them to be a new proposal a_s.
13. Assume $l := l + 1$ and go to 1.

5 Numerical Example

To illustrate our procedure let us consider the project selection problem. Ten proposals are evaluated with respect to four attributes. The evaluations of alternatives with respect to attributes are presented in Table 3. We assume that the DM is risk-averse. To identify the first proposal, stochastic dominance relations were identified (Table 4).

The first proposal is the alternative a_6, as $d_{61} = 5$, $d_{62} = 5$, $d_{63} = 3$, $d_{64} = 5$, and $\overline{d}_6 = 5$. We assume: $l = 1$,

$$\mathbf{A}^{(0)} = \mathbf{A} = \{a_1, a_2, a_3, a_4, a_5, a_6, a_7, a_8, a_9, a_{10}\}.$$

Iteration 1:

1. The DM decides that for each attribute means should be presented in the dialog phase of the procedure.
2. The data are presented to the decision maker (Table 5).
3. The DM is not satisfied with the proposal.
4. The DM would like to improve the evaluation with respect to attribute X_2.
5. The DM sets the order of other attributes starting from the one that can be decreased first: X_3, X_4, X_1.
6. The alternatives with evaluations that are not dominated by the evaluation of alternative a_6 with respect to attribute X_2 are identified:

$$\mathbf{A}^{(1)} = \{a_1, a_4, a_5, a_7, a_9, a_{10}\}$$

7. To identify a new proposal we analyze the relations between point-to-point trade-offs for pairs of attributes: (X_2, X_3), (X_2, X_4), (X_2, X_1). The set of potential new proposals is:

$$\mathbf{B} = \mathbf{A}^{(1)} = \{a_1, a_4, a_5, a_7, a_9, a_{10}\}.$$

8. We start to analyze point-to-point trade-offs with the pair of attributes (X_2, X_3). We generate distributions of trade-offs for each pair (a_i, a_6) such that $a_i \in \mathbf{A}^{(1)}$ and analyze SD relations (Table 6).
9. As distributions of trade-offs for the pairs (a_4, a_6), (a_5, a_6), (a_9, a_6), (a_{10}, a_6) are dominated, the alternatives a_4, a_5, a_9 and a_{10} are excluded form the set of potential new proposals:

$$\mathbf{B} = \mathbf{B} \backslash \{a_4, a_5, a_9, a_{10}\} = \{a_1, a_7\}.$$

10. As the set **B** consists of more than one alternative, we analyze relations between trade-offs distributions for the next pair of attributes (X_2, X_4). Unfortunately no SD relations can be identified for this pair of alternatives. The same situation is for attributes (X_2, X_1).

Table 3 Evaluations of alternatives with respect to attributes

X_1	Projects									
	1	2	3	4	5	6	7	8	9	10
1						1/7		1/7		
2	3/7						1/7	2/7		1/7
3	1/7				1/7			2/7		2/7
4							2/7	1/7		2/7
5	2/7		3/7	1/7		3/7	1/7	1/7	2/7	1/7
6		1/7	1/7		2/7	1/7	2/7		1/7	
7	1/7	2/7	1/7	1/7	2/7				3/7	1/7
8		2/7	2/7	1/7		1/7	1/7		1/7	
9				3/7	2/7					
10		2/7		1/7		1/7				

X_2	Projects									
	1	2	3	4	5	6	7	8	9	10
1		1/7						3/7		
2		3/7	2/7					3/7		1/7
3		1/7	1/7	1/7		4/7			1/7	
4				1/7				1/7	1/7	
5		1/7	2/7		1/7					
6	1/7	1/7		1/7	2/7		1/7		1/7	
7					1/7	1/7	1/7		4/7	2/7
8	2/7		1/7	3/7	2/7	2/7	3/7			3/7
9	3/7		1/7	1/7	1/7		2/7			
10	1/7									1/7

X_3	Projects									
	1	2	3	4	5	6	7	8	9	10
1		2/7								1/7
2		1/7			3/7					2/7
3	1/7	4/7		1/7	1/7				1/7	
4	3/7				1/7	1/7			2/7	
5					2/7	1/7		1/7	2/7	
6	1/7									2/7
7				1/7			2/7	1/7	2/7	2/7
8	1/7		4/7	2/7		2/7	3/7	2/7		
9	1/7		3/7	1/7		1/7	1/7	3/7		
10				2/7		2/7	1/7			

X_4	Projects									
	1	2	3	4	5	6	7	8	9	10
1				2/7						
2									1/7	
3	3/7						1/7			
4				1/7					1/7	
5	2/7			2/7		1/7	1/7			
6				1/7	1/7	1/7			3/7	3/7
7			1/7		1/7	1/7				1/7
8	1/7	4/7	4/7	1/7	3/7	2/7	3/7	3/7	1/7	1/7
9		2/7			1/7	1/7	1/7	1/7		1/7
10	1/7	1/7	2/7		1/7	1/7	1/7	3/7	1/7	1/7

Table 4 Stochastic dominance relations between distributional evaluations

X_1	Projects									
	a_1	a_2	a_3	a_4	a_5	a_6	a_7	a_8	a_9	a_{10}
a_1								FSD		
a_2	FSD		FSD		FSD	FSD	FSD	FSD	FSD	FSD
a_3	FSD					SSD	FSD	FSD		FSD
a_4	FSD		FSD		FSD	FSD	FSD	FSD	FSD	FSD
a_5	FSD					SSD	FSD	FSD		FSD
a_6								FSD		
a_7	FSD							FSD		FSD
a_8										
a_9	FSD		SSD			SSD	FSD	FSD		FSD
a_{10}	SSD							FSD		

X_2	Projects									
	a_1	a_2	a_3	a_4	a_5	a_6	a_7	a_8	a_9	a_{10}
a_1		FSD	FSD	FSD	FSD	FSD	FSD	FSD	FSD	FSD
a_2								FSD		
a_3		FSD						FSD		
a_4		FSD	FSD			FSD		FSD	FSD	
a_5		FSD	FSD	SSD		FSD		FSD	FSD	
a_6		FSD	SSD					FSD		
a_7		FSD	FSD	FSD	FSD	FSD		FSD	FSD	SSD
a_8										
a_9		FSD	SSD			SSD		FSD		
a_{10}		FSD	FSD					FSD		

X_3	Projects									
	a_1	a_2	a_3	a_4	a_5	a_6	a_7	a_8	a_9	a_{10}
a_1		FSD			FSD					SSD
a_2										
a_3	FSD	FSD		SSD	FSD	SSD	SSD	FSD	FSD	FSD
a_4	FSD	FSD			FSD				FSD	FSD
a_5		FSD								
a_6	FSD	FSD			FSD				FSD	FSD
a_7	FSD	FSD		SSD	FSD	SSD		SSD	FSD	FSD
a_8	FSD	FSD		SSD	FSD	SSD			FSD	FSD
a_9		FSD			FSD					SSD
a_{10}		FSD								

X_4	Projects									
	a_1	a_2	a_3	a_4	a_5	a_6	a_7	a_8	a_9	a_{10}
a_1				SSD						
a_2	FSD		SSD	FSD	FSD	FSD	FSD		FSD	FSD
a_3	FSD			FSD	FSD	FSD	FSD		FSD	FSD
a_4										
a_5	FSD			FSD		FSD	FSD		FSD	FSD
a_6	FSD			FSD			SSD		FSD	
a_7	FSD			FSD					FSD	
a_8	FSD	FSD	FSD	FSD	FSD	FSD	FSD		FSD	FSD
a_9				FSD						
a_{10}	FSD			FSD			SSD		FSD	

Table 5 Data presented to the DM at iteration 1

	Mean			
	X^1	X^2	X^3	X^4
a_6	5,714	5,000	7,714	7,571
Max	8,143	8,429	8,429	9,000

Table 6 Iteration 1 – SD relations between trade-offs for attributes (X_2, X_3)

	T_{16}^{23}	T_{46}^{23}	T_{56}^{23}	T_{76}^{23}	T_{96}^{23}	T_{106}^{23}
T_{16}^{23}		SSD	FSD		FSD	FSD
T_{46}^{23}			SSD		FSD	FSD
T_{56}^{23}					SSD	
T_{76}^{23}		FSD	FSD		FSD	FSD
T_{96}^{23}						
T_{106}^{23}					SSD	

11. As relations between trade-offs for each pair of attributes have been analyzed and the set of potential new proposals still consists of more than one alternative, we analyze relations between the alternatives a_4 and a_7 with respect to the attribute that should be improved, that is, X_2. As $X_1^2 \succ_{\text{FSD}} X_7^2$, we assume the alternative a_1 to be a new proposal.

The procedure is continued in the same way, until the DM is satisfied with the proposal.

6 Conclusions

In many cases, the DM faced with a candidate solution is able to answer the simplest questions only: which attribute should be improved and which attributes can be decreased. In such a situation trade-offs can be used for generation of a new proposal. When the evaluations of alternatives with respect to attributes are characterized by random variables, a point-to-point trade-off is characterized by a random variable as well.

In this paper a new interactive procedure based on the treatment of trade-offs has been proposed. The procedure requires a limited amount of preference information from the DM.

The procedure presented in this work can also be applied for mixed problems, i.e. problems in which evaluations with respect to some attributes take the form of probability distributions, while the remaining ones are deterministic.

The proposed technique may be useful for various types of problems in which uncertain outcomes are compared. It has been designed for problems with up to moderate number of discrete alternatives (not more than hundreds) and can be applied in such areas as, for example, inventory models, evaluation of investment projects, production process control, and many others.

Acknowledgements The research has been supported by Polish Ministry of Science and Higher Education under project NN111 235036 (support in years 2009–2012).

References

Benayoun R, de Montgolfier J, Tergny J, Larichev C (1971) Linear programming with multiple objective functions: step method (STEM). Math Program 8:366–375

D'Avignon G, Vincke Ph (1988) An outranking method under uncertainty. Eur J Oper Res 36: 311–321

Dendrou BA, Dendrou SA, Houtis EN (1980) Multiobjective decisions analysis for engineering systems. Comput Oper Res 7:301–312

Geoffrion AM, Dyer JS, Feinberg A (1972) An interactive approach for multi-criterion optimization with an application to the operation of an academic department. Manag Sci 19:357–368

Huang CC, Kira D, Vertinsky I (1978) Stochastic dominance rules for multiattribute utility functions. Rev Econ Stud 41:611–616

Kaliszewski I, Michalowski W (1999) Searching for psychologically stable solutions of multiple criteria decision problems. Eur J Oper Res 118:549–562

Keeney RL, Raiffa H (1976) Decisions with multiple objectives: preferences and value tradeoffs. Wiley, New York

Martel JM, D'Avignon G, Couillard J (1986) A fuzzy relation in multicriteria decision making. Eur J Oper Res 25:258–271

Nowak M (2004) Interactive approach in multicriteria analysis based on stochastic dominance. Contr Cybern 33:463–476

Nowak M (2006) INSDECM – an interactive procedure for stochastic multicriteria decision problems. Eur J Oper Res 175:1413–1430

Ogryczak W, Ruszczyski A (1999) Form stochastic dominance to mean-risk models: semideviations as risk measures. Eur J Oper Res 116:33–50

Saaty TL, Vargas LG (1987) Uncertainty and rank order in the analytic hierarchy process. Eur J Oper Res 32:107–117

Spronk J (1981) Interactive multiple goal programming. Martinus Nijhoff, The Hague

Steuer RE (1986) Multiple criteria optimization: theory, computation and application. Wiley, New York

Wierzbicki A (1980) The use of reference objectives in multiobjective optimization. In: Fandel G, Gal T (eds) MCDM theory and application. Springer, Berlin, pp 468–486

Zaras K, Martel JM (1994) Multiattribute analysis based on stochastic dominance. In: Munier B, Machina MJ (eds) Models and experiments in risk and rationality. Kluwer Academic, Dordrecht, pp 225–248

Zionts S, Wallenius J (1976) An interactive programming method for solving the multiple criteria problem. Manag Sci 22:652–663

Interactive Multiobjective Optimization for 3D HDR Brachytherapy Applying IND-NIMBUS

Henri Ruotsalainen, Kaisa Miettinen, and Jan-Erik Palmgren

Abstract An anatomy based three-dimensional dose optimization approach for HDR brachytherapy using interactive multiobjective optimization is presented in this paper. In brachytherapy, the goals are to irradiate a tumor without causing damage to healthy tissue. These goals are often conflicting, i.e. when one target is optimized the other one will suffer, and the solution is a compromise between them. Our interactive approach is capable of handling multiple and strongly conflicting objectives in a convenient way, and thus, the weaknesses of widely used optimization techniques (e.g. defining weights, computational burden and trial-and-error planning) can be avoided. In addition, our approach offers an easy way to navigate among the obtained Pareto optimal solutions (i.e. different treatment plans), and plan quality can be improved by finding advantageous trade-offs between the solutions. To demonstrate the advantages of our interactive approach, a clinical example of seeking dwell time values of a source in a gynecologic cervix cancer treatment is presented.

1 Introduction

Radiation's delivery in high-dose-rate (HDR) intracavitary brachytherapy using an afterloading unit is realized by using temporarily implanted catheters: a programmable remote unit moves a single radioactive source along catheters. This system produces a high-dose region centered on the planning target volume while sparing the adjacent bladder and bowel. The flexibility of this system allows it to be tailored to a variety of different patient anatomy and cancer types because a wide variety of dose distributions can be generated from a given implant simply by adjusting the length of time (dwell time) that the source dwells at any location within a catheter (dwell position). In clinics, this flexibility allows the full benefit of the use

---

H. Ruotsalainen (✉)
Department of Physics, University of Kuopio, P.O. Box 1627, 70211 Kuopio, Finland
e-mail: Henri.Ruotsalainen@uku.fi

D. Jones et al. (eds.), *New Developments in Multiple Objective and Goal Programming*, Lecture Notes in Economics and Mathematical Systems 638,
DOI 10.1007/978-3-642-10354-4_8,

of three-dimensional (3D) planning system based on computer tomography (CT) or magnetic resonance imaging (MRI).

However, the increased flexibility in treatment applications and imaging increases also the complexity in the treatment planning. A patient domain can be divided into three different parts based on a patient anatomy: a planning target volume (PTV), dose sensitive organs at risk (OARs) and healthy normal tissue (NT). The OARs and NT are typically near the PTV, and thus, they may be unnecessarily overdosed. To maintain a complete coverage of the PTV and simultaneously reduce the dose to NT and OARs, the dose distribution should be as conformal as possible to the relevant anatomy.

Recently, there has been interest in using multiobjective optimization in brachytherapy treatment planning (e.g. Yu, 1997; Lahanas et al., 1999; Yu et al., 2000; Lessard et al., 2006). This is because the aim of brachytherapy is to treat the tumor without affecting healthy tissue but, naturally, increasing the dose in the tumor also increases the unwanted dose in surrounding healthy tissue. Thus, when one target is optimized, the other will suffer, and the solution is a compromise between them. This trade-off is complex, and optimization tools capable of handling multiple and conflicting objectives are naturally required. The multiobjective optimization approaches presented in the literature are based on using objective weights defined beforehand, where the final objective function is expressed as a weighted sum of the conflicting objectives (e.g. Milickovic et al., 2002; Lahanas and Baltas, 2003). In these cases, objectives are often formulated as using penalties where exceeding predefined upper limits for doses are penalized (e.g. Lahanas et al., 2003b). Unfortunately, it is typically hard to predefine the priorities or weights of the optimization targets. Moreover, sometimes information about objectives and even the practical relevance of the objective functions can become blurred if the objectives are expressed as a sum. Furthermore, penalizing only the overdose should not be the actual goal as we argue in this paper. Alternatively, evolutionary algorithms (e.g. Lahanas et al., 2001; Milickovic et al., 2001) have been used, too. These methods have their own difficulties because they are very time consuming requiring a lot of calculation when computing a large set of approximating solutions. The similar optimization methods and methods for comparing different solutions have been studied also in a similar context in intensity modulated radiotherapy (IMRT) treatment planning in Romeijn et al. (2004), Craft et al. (2005), Hoffmann et al. (2006), Holder (2006), Craft et al. (2007), Thieke et al. (2007), Monz et al. (2008), Craft and Bortfeld (2008) and Ehrgott and Winz (2008).

To overcome some shortcomings of currently used approaches, we exploit an interactive multiobjective optimization method for 3D HDR brachytherapy optimization in this paper. The real multiobjective nature of the problem is taken into account in the problem formulation and in the interactive solution process. For some reason, interactive multiobjective optimization methods have not been studied in the field of brachytherapy optimization before. The studies where brachytherapy treatment plan has been optimized are based on a priori methods or a posteriori methods. However, according to our knowledge, an interactive multiobjective optimization method is ideal for brachytherapy optimization, and we demonstrate the advantages of our interactive approach by an example of a treatment plan of cervical cancer. In this study, our interactive approach is used to determine the dwell time values needed to fulfill the prescribed dose to the tumor and to minimize dose in each organ at risk. In our approach, the decision maker's (i.e. treatment planner's) knowledge and preferences are used during the iterative optimization process to direct the search in order to find the most preferred plan, that is, the best Pareto optimal solution, as it is called, between the conflicting treatment planning targets. This makes treatment planning times shorter, and good trade-offs between the targets can be found to improve the treatment plan's quality. Furthermore, the interactive approach improves the decision makers control over treatment: with this system, the treatment planner plays directly with the compromises between target coverage and protection of organs at risk instead of with dwell positions, dwell times, and objective weights. This approach brings the planning process near to the real clinical issues avoiding artificial simplifications, and when compared to the currently used trial-and-error method, our approach guarantees the mathematical optimality of the final solution, i.e. treatment plan. Here, by mathematical optimality we refer to Pareto optimality which means that any of the targets cannot be improved without impairing at least one other target at the same time. Pareto optimality is not guaranteed by trial-and-error method used at the clinics in which some of the targets could still be improved without deteriorating other targets. It is important to point out that these kinds of tools are designed to assist human treatment planners in their work, not to replace them.

2 Methods

2.1 Dose Calculation

Before optimization, the dose distribution in a patient needs to be calculated. The dose $D(x_i) = D_i$ at the ith sampling point x_i is calculated by

$$D_i = \sum_{j=1}^{p} t_j d_{ij}, \tag{1}$$

where p is the number of sources, t_j is the dwell time of the jth source dwell position and d_{ij} is the kernel value, i.e. dose value, for the ith dose calculation point and jth source dwell position. The dose rate matrix d_{ij} can be calculated using the following equation according to TG43 (Nath et al., 1995; Rivard et al., 2004):

$$d_{ij} = S_k \Lambda \Phi_{an}(\theta, r_{ij}) g(r_{ij}) / r_{ij}^2, \tag{2}$$

where S_k is the air kerma strength, Λ is the dose-rate constant, $\Phi_{an}(\theta, r_{ij})$ is the anisotropy function, $g(r_{ij})$ is the radial dose functions, and r_{ij} the distance between the dwell position j and the dose calculation point i (point source). In this paper, we use (1) for dose calculations (anisotropy of a patient is neglected). In interactive multiobjective optimization, the dwell times t_j are the decision variables.

2.2 Objective Function Formulation

The aim of brachytherapy treatment planning is to obtain a plan which covers the PTV with at least some specified dose value D_{PTV}, which is case-specific depending on the type of the tumor. In addition to this, there is an upper bound for the dose in NT and the OAR which should not be exceeded. We denote these bounds by D_{NT} and D_{OAR}. Traditionally in optimization, dwell times t are sought so that the above-mentioned requirements are fulfilled, that is

$$\begin{aligned} D_i &\geq D_{\mathrm{PTV}},\ i \in I_{\mathrm{PTV}}, \\ D_i &\leq D_{\mathrm{NT}},\ \ i \in I_{\mathrm{NT}}, \\ D_i &\leq D_{\mathrm{OAR}},\ i \in I_{\mathrm{OAR}}, \end{aligned} \tag{3}$$

where I_{PTV}, I_{NT} and I_{OAR} present indexes of sampling points located in a region PTV, NT and OAR, respectively. In the literature, several different objective functions have been used to fulfill these requirements; variance based objective functions (e.g. Lahanas et al., 2003a) or dose volume histogram based objective functions (e.g. Lahanas et al., 2003b), for example. In addition, the formulation used in Lessard and Pouliot (2001) and Lessard et al. (2002) is well known. However, let us point out that even though (3) describes an acceptable solution, it is important to carefully think what should actually be optimized: the goal is that the dose in NT and OAR should be as low as possible (minimized), not only under the predefined bounds.

Now, based on the fact that we want to minimize the dose in NT and OAR (i.e. not only the dose exceeding limits D_{NT} and D_{OAR}), objective functions can be formulated (in a discrete form) as

$$f_1(t) = \frac{|I_{\tilde{\mathrm{PTV}}}|}{|I_{\mathrm{PTV}}|}, \tag{4}$$

$$f_2(t) = \frac{1}{|I_{\mathrm{NT}}|} \sum_{i \in I_{\mathrm{NT}}} D_i, \tag{5}$$

$$f_3(t) = \frac{1}{|I_{\mathrm{OAR}}|} \sum_{i \in I_{\mathrm{OAR}}} D_i \tag{6}$$

and

$$f_4(t) = \max_{i \in I_{\mathrm{PTV}}} D_i, \tag{7}$$

where t is a vector of dwell time values, and $|I_{\mathrm{PTV}}|$, $|I_{\mathrm{OAR}}|$ and $|I_{\mathrm{NT}}|$ denote number of sampling points, i.e. size of the set, in a region (PTV, OAR and NT). Because of the computational reasons, all the sampling points are situated on the surface of the region, and thus a maximum dose inside the PTV is not an objective but it is controlled later. Here, $|I_{\tilde{\mathrm{PTV}}}|$ represents the number of sampling points in the PTV that have a dose value bigger than the dose limit D_{PTV}. Thus, the function f_1 represents a percentual volume where the dose is higher or equally high to the prescribed dose D_{PTV} in the PTV, and it is maximized. The functions f_2 and f_3 are the averaged doses on the surface of NT and the OAR, respectively, to be minimized. If there are multiple OARs (as also here in the example), there are as many objective functions each similar to f_3. The objective function f_4 describes the maximum dose on the surface between the PTV and NT (because the sampling points are situated on the surface of the PTV). With these objective functions, the unwanted dose in NT and the OAR is really minimized, not only penalized if it exceeds predefined upper limits for doses as it is often presented in the literature, see Lahanas et al. (2003b), for example.

2.3 Multiobjective Optimization

2.3.1 Multiobjective Optimization Problem

A multiobjective optimization problem can be defined as follows (Miettinen, 1999)

$$\begin{aligned} &\text{minimize } \{f_1(t), f_2(t), \ldots, f_k(t)\} \\ &\text{subject to } t \in S, \end{aligned} \tag{8}$$

where t is a vector of decision variables from the feasible set $S \subset \mathbf{R}^n$ defined by linear, nonlinear and box constraints. We can denote an objective vector by $\mathbf{f}(t) = (f_1(t), f_2(t), \ldots, f_k(t))^T$. Furthermore, we denote the image of the feasible set by $\mathbf{f}(S) = Z$ and call it a feasible objective set. In multiobjective optimization, optimality is understood in the sense of Pareto optimality (Miettinen, 1999). A decision vector $t' \in S$ is Pareto optimal if there does not exist another decision vector $t \in S$ such that $f_i(t) \leq f_i(t')$ for all $i = 1, \ldots, k$ and $f_j(t) < f_j(t')$ for at least one index j. These Pareto optimal solutions form a Pareto optimal set. All the solutions are equally good from a mathematical point of view, and they can be regarded as equally valid compromise solutions of the problem. There exists no trivial mathematical tool in order to find the best solution in the Pareto optimal set because vectors cannot be ordered completely. That is why we need some additional information.

Typically, a decision maker, who is an expert in the field from where the problem has arisen (here, a treatment planner), is needed in order to find the best or most satisfying solution, called the final one. The decision maker can participate in the solution process, and, in one way or another, determine which of the Pareto optimal solutions is the most satisfying to be the final solution. It can be useful for the decision maker to know the ranges of objective function values in the Pareto optimal set. An ideal objective vector $\mathbf{z}^* \in \mathbf{R}^k$ gives lower bounds for the objective functions in the Pareto optimal set and it is obtained by minimizing each objective function individually subject to the constraints. A nadir objective vector $\mathbf{z}^{\mathrm{nad}}$ giving upper bounds of objective function values in the Pareto optimal set is usually difficult to calculate, and thus its values are usually only approximated by using pay-off tables, for example (see Miettinen, 1999 for details).

According to Miettinen (1999), the methods developed for multiobjective optimization can be divided into four classes depending on the role of the decision maker. There are methods for use when no decision maker is available. In these methods, the final solution is some neutral compromise solution. In the three other classes, the decision maker participates in the solution process beforehand, afterwards or iteratively: these methods are called the a priori, a posteriori and interactive methods, respectively. It can be difficult for the decision maker to specify preferences before the solution process has started and, on the other hand, generating many Pareto optimal solutions for the decision maker to compare can be computationally costly. It is also problematic to compare many solutions without setting too much cognitive load on the decision maker. Consequently, and encouraged by experiences reported in Ruotsalainen et al. (2006), we concentrate in this paper on interactive methods.

2.3.2 The Interactive Multiobjective Optimization Method NIMBUS

In this paper, we integrate an anatomy based 3D HDR brachytherapy dose calculation model with an interactive multiobjective optimization method. The method we use is NIMBUS (Miettinen, 1999; Miettinen and Mäkelä, 2006, 1995). This method has been successfully used in external radiotherapy treatment planning optimization in an academic case with a simple pencil beam model in Ruotsalainen et al. (2006).

In interactive multiobjective optimization methods, the information given to and required from the decision maker must be easily understandable. The NIMBUS method is based on the idea of classification of objective functions. It is known that classification can be considered an acceptable task for human decision makers from a cognitive point of view (Larichev, 1992). In NIMBUS, the decision maker participates in the solution process iteratively and continuously. Finally, he/she decides which of the Pareto optimal solutions obtained is the most desired one. During the solution process, the decision maker classifies objective functions at the current Pareto optimal point into up to five classes. The classes are the following:

- I^{imp} functions whose values should be improved
- I^{asp} functions whose values should be improved up to a desired aspiration level $\hat{z}$
- I^{sat} functions whose values are satisfactory
- I^{bound} functions whose values can be impaired up to a given bound ϵ
- I^{free} functions whose values can change freely

Since all the solutions considered are Pareto optimal, the decision maker cannot make a classification where all the objective function values should improve without allowing at least one of the objective functions to be impaired. The aspiration levels and the bounds are elicited from the decision maker during the classification procedure if they are needed.

By classifying the objective functions, the decision maker gives preference information about how the current solution should be improved. Based on that, a scalarized single objective optimization problem (a subproblem, as we call it) can be formed, and it can be solved with an appropriate solver. Here, we use a synchronous NIMBUS method (Miettinen and Mäkelä, 2006). In this method, there are four different subproblems available, so the decision maker can choose whether to see one to four new solutions after each classification. Each subproblem generates a new Pareto optimal solution that satisfies the preferences given in the classification as well as possible, but the preferences are taken into account in slightly different ways (Miettinen and Mäkelä, 2002). The decision maker can use any solution obtained so far as a starting point for a new classification, and interesting solutions can also be saved in a database, so that the solution process can be continued later from any of them. Alternatively, the decision maker can generate a desired number of Pareto optimal intermediate solutions between any two solutions. This capacity differs from many other approaches used in treatment planning where intermediate solutions are only approximated, see, e.g. Monz et al. (2008). For more information about the NIMBUS algorithm, the scalarizations used and ways of aiding comparison of Pareto optimal solutions generated with different visualizations, see Miettinen and Mäkelä (2006).

3 Results

3.1 Problem Settings

Here, a clinical example of seeking dwell time values of a source in a gynecologic cervix cancer treatment is presented. In the example, Fletcher–Suit intracavitary applicator system was used to deliver the radiation, and there were 17 possible dwell positions (resolution of 5 mm in three applicators). Thus, the number of continuous decision variables was 17. In addition, the number of sampling points in computations was 508. The problem contained box constraints for the decision variables (i.e. dwell times). In the example, there were two OARs (bladder and rectum, sigma was not adjacent to the tumor), and thus, there were two objective functions similar to f_3 (f_3^{bladder} and f_3^{rectum}).

In the example, all the simulations were carried out with the mathematical software Matlab® R2006b after the patient geometry (anatomy and sampling points) was generated with a treatment planning software (BrachyVision®, Varian Medical Systems, software version 7.3.10) at the Kuopio University Hospital. The optimization was done with a personal computer (Pentium® 4 CPU 3.00 GHz with

2 GB central memory). For interactive multiobjective optimization, an implementation of the NIMBUS method, called IND-NIMBUS® was used Miettinen (2006). A global optimization method (computation time was minutes per classification with the presented PC) was used to solve the formed subproblem in IND-NIMBUS. This optimization method does not require continuity of the objective functions.

3.2 *Fletcher–Suit Applicator Example*

The optimization problem used for the demonstration of the proposed interactive multiobjective optimization approach has the form

$$\begin{aligned} &\text{optimize } \{f_1(t), f_2(t), f_3^{\text{bladder}}(t), f_3^{\text{rectum}}(t), f_4(t)\} \\ &\text{subject to } t \in S, \end{aligned} \tag{9}$$

where t is a vector of continuous decision variables, and $S = [0, 100] \times [0, 100] \times \cdots \times [0, 100] \subset \mathbf{R}^{17}$. Objective functions $f_1 - f_4$ are defined in Sect. 2.2: the value of f_1 represents the percentual value of sampling points in the PTV which has a dose value higher than the dose limit D_{PTV} (7 Gy), and values f_2, f_3^{bladder}, f_3^{rectum} represent averaged dose values in NT, bladder, and rectum, respectively (in gray). Finally, f_4 is the maximum dose on the surface between the PTV and NT (in gray).

3.2.1 Interactive Solution Process

The interactive solution process (i.e. moving from one Pareto optimal solution to another) was guided by preference information of a treatment planner, who was acting as a decision maker. Before the solution process, the decision maker had the following desires: the percentual value of sampling points in the PTV that have a dose value higher than the dose limit D_{PTV} should be maximized (f_1). At the same time, the averaged doses in NT and both OARs should be minimized (f_2, f_3^{bladder}, f_3^{rectum}). Since the objective function f_4 shows the maximum dose on the surface between the PTV and NT, the decision maker wanted to minimize it, too. At the very beginning, the objective functions had the initial values (generated by IND-NIMBUS) $f_1 = 0.58$, $f_2 = 9.89$, $f_3^{\text{bladder}} = 1.96$, $f_3^{\text{rectum}} = 3.13$ and $f_4 = 96.80$ (initial solution $\mathbf{f}(t^1)$). As can be seen from the initial objective function values $\mathbf{f}(t^1)$, the f_1 value was certainly too low ($f_1 = 0.58$, i.e. 58% of the PTV received higher dose than D_{PTV} which was 7 Gy). Nevertheless, the objective functions f_2, f_3^{bladder} and f_3^{rectum} were in a good level and, thus, the dose in NT and the OARs was low but, as said, at the same time the dose in the PTV was too low and the tumor would not be treated properly. Thus, the decision maker wanted to search for a better solution in an iterative way. He started to classify the functions and generated new solutions (see classes in Sect. 2.3.2), and in this way declared his preferences and steered the solution process interactively and iteratively towards the most satisfying Pareto optimal solutions.

Table 1 Summary of interactive solution process. Bounds and aspiration levels used are denoted as superscripts in the classification notation

Solution	$f_1(-)$	f_2(Gy)	f_3^{bladder}(Gy)	f_3^{rectum}(Gy)	f_4(Gy)
Ideal	1	0	0	0	0
Nadir	0	25.82	4.64	7.72	744.33
Initial solution					
$\mathbf{f}(t^1)$	**0.58**	**9.89**	**1.96**	**3.13**	**96.80**
1st classification	$I^{\text{asp } 0.70}$	I^{free}	$I^{\text{bound } 2.00}$	$I^{\text{asp } 2.00}$	I^{free}
$\mathbf{f}(t^2)$	**0.58**	**9.55**	**1.98**	**2.74**	**135.89**
2nd classification	I^{imp}	I^{free}	$I^{\text{bound } 2.00}$	$I^{\text{asp } 2.00}$	I^{free}
$\mathbf{f}(t^3)$	0.83	11.18	2.67	3.29	55.18
$\mathbf{f}(t^4)$	1.00	15.86	3.34	4.05	109.89
$\mathbf{f}(t^5)$	0.79	11.57	2.55	3.24	76.85
Intermediate sol.[a]					
$\mathbf{f}(t^6)$	0.64	9.87	2.12	2.86	64.23
$\mathbf{f}(t^7)$	0.69	10.02	2.24	2.90	62.11
$\mathbf{f}(t^8)$	0.71	10.44	2.36	3.04	52.23
$\mathbf{f}(t^9)$	**0.73**	**10.50**	**2.35**	**3.15**	**52.23**
$\mathbf{f}(t^{10})$	0.76	10.87	2.45	3.22	54.00

[a] Intermediate solutions between $\mathbf{f}(t^2)$ and $\mathbf{f}(t^3)$

In the *1st classification*, as said, he wanted to obtain a better value to f_1 (aspiration level 0.70) and simultaneously maintain the good values of f_3 and f_4 (save the OARs). Thus, he set a bound to f_3^{bladder}, and an aspiration level to f_3^{rectum}. The bound and aspiration level both were 2.00. At the same time, he had to allow some other targets (f_2 and f_4) to get worse. That is, the classification was $f_1 : I^{\text{asp } 0.70}$, $f_2 : I^{\text{free}}$, $f_3^{\text{bladder}} : I^{\text{bound } 2.00}$, $f_3^{\text{rectum}} : I^{\text{asp } 2.00}$ and $f_4 : I^{\text{free}}$, and he wanted to generate one new solution (solutions are collected in Table 1). After the first classification, the decision maker obtained a better solution ($\mathbf{f}(t^2)$) because the rectum ($f_3^{\text{rectum}} = 2.74$), which he considered very important, obtained smaller dose value than in solution $\mathbf{f}(t^1)$, but still f_1 was too low according to his preferences. Because of this reason, he decided to do a *2nd classification* using the solution $\mathbf{f}(t^2)$ as a starting point of the classification. In this classification, the decision maker wanted to improve f_1 as much as possible, and again, he set a bound to f_3^{bladder}, and an aspiration level to f_3^{rectum} to maintain the good levels of these objectives. In addition, the decision maker allowed f_2 and f_4 to change freely. Therefore, the classification was $f_1 : I^{\text{imp}}$, $f_2 : I^{\text{free}}$, $f_3^{\text{bladder}} : I^{\text{bound } 2.00}$, $f_3^{\text{rectum}} : I^{\text{asp } 2.00}$ and $f_4 : I^{\text{free}}$. After the second classification, the decision maker obtained three new solutions having excellent f_1 values, but, at the same time, values of other objectives were not so good (Table 1). That is why he wanted to generate five *intermediate solutions* between the solutions $\mathbf{f}(t^2)$ and $\mathbf{f}(t^3)$, which had good values of objectives f_3^{bladder} and f_3^{rectum}, and f_1, respectively. Intermediate solutions represent compromise solutions between the conflicting treatment planning targets, and the decision maker was able to choose the best Pareto optimal solution according to his knowledge to be the final solution, i.e. the final treatment plan. That solution was $\mathbf{f}(t^9)$, in which the objective values were $f_1 = 0.73$, $f_2 = 10.50$, $f_3^{\text{bladder}} = 2.35$, $f_3^{\text{rectum}} = 3.15$ and $f_4 = 52.23$. As can be seen in Figs. 1 (left) and 2, all the requirements of the

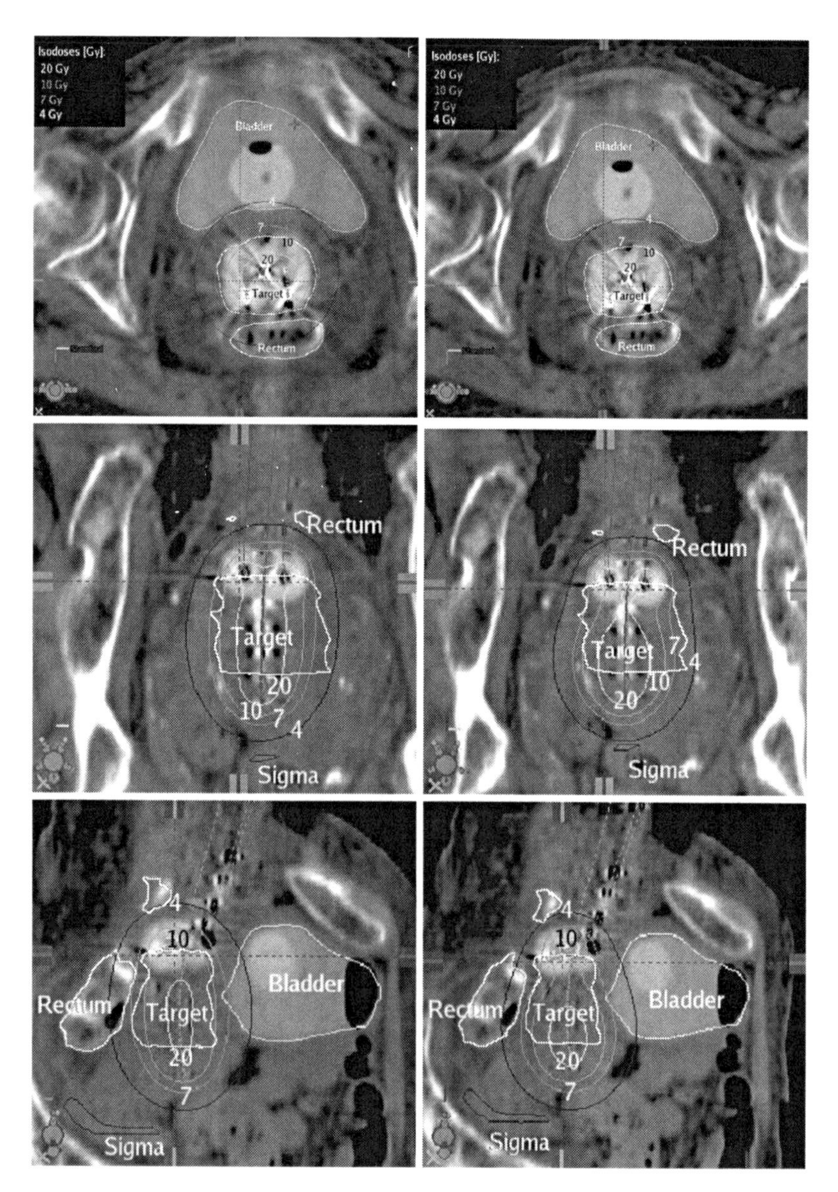

Fig. 1 *Left*, the final dose distribution with interactive multiobjective optimization (solution $\mathbf{f}(t^9)$) from three different point of views (x, y, and z-direction), and *right*, for comparison, solution obtained with BrachyVision® optimization

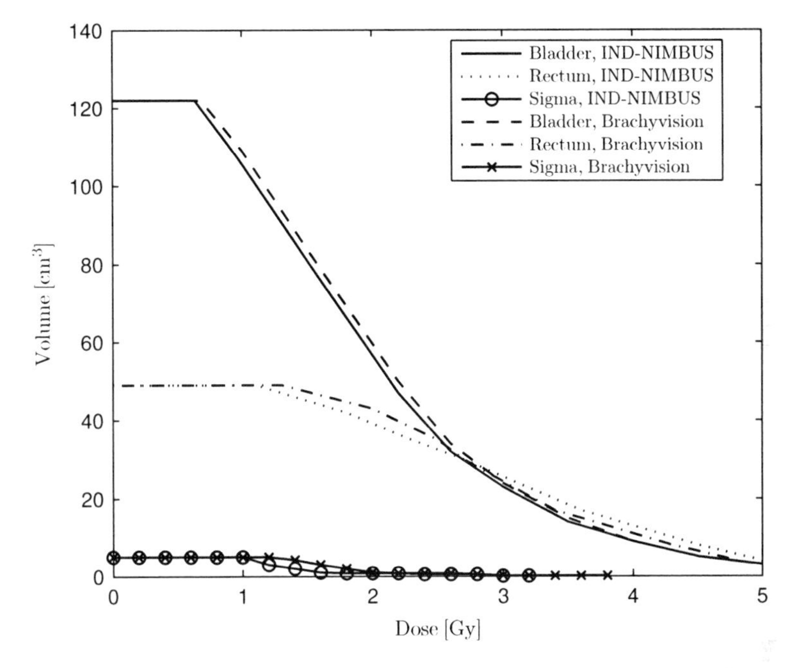

Fig. 2 For comparison, dose volume histograms of OARs. Dose volume histogram values start to decrease faster in IND-NIMBUS solution than in solution obtained with Brachyvision®

treatment plan were taken into account as well as possible: harmful dose in rectum and bladder was minimized and the prescribed dose in the PTV was delivered. Thus, the treatment plan was clinically acceptable.

All the solutions obtained and steps taken by the decision maker during the solution process are collected in Table 1. In this table, the starting point of a new classification and the final solution are given in bold face. Let us add that a more thorough description of a typical process of classifying objective functions in a radiotherapy case and steering the optimization process is presented in Ruotsalainen et al. (2006).

3.2.2 Comparison and Discussion

In this example, we have shown how our interactive approach can handle the strongly conflicting objective functions in a cervix cancer case. As can be seen in Fig. 3 (a display of IND-NIMBUS software), the solutions obtained can be compared and carefully studied during the interactive solution process. Thus, the decision maker is better prepared to make the final decision, i.e. choose the final treatment plan, after analyzing the isodose maps (Fig. 1), dose volume histograms (Fig. 2), graphical information (Fig. 3), and numerical information (Table 1 and

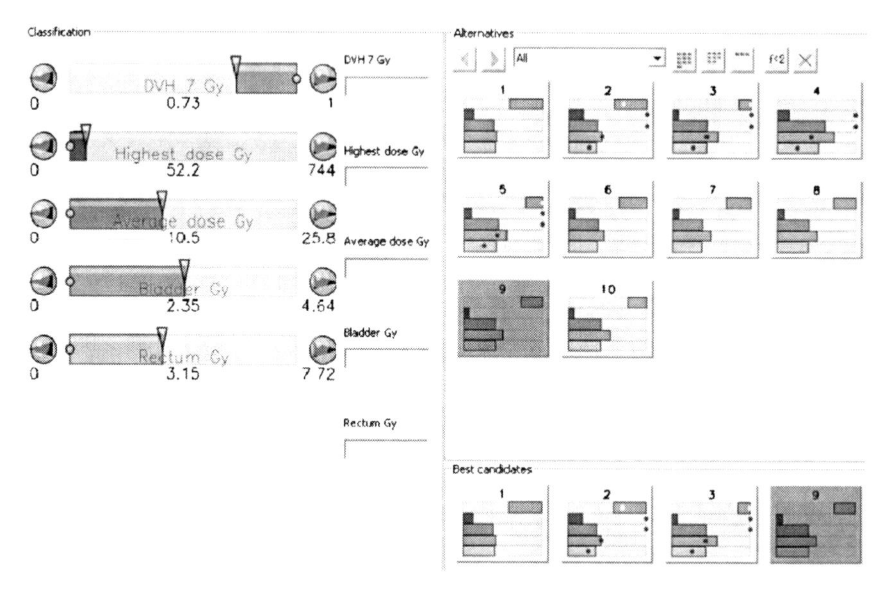

Fig. 3 All the solution obtained during the optimization process can be compared also with a graphical tool in IND-NIMBUS. On the *left*, the final solution $\mathbf{f}(t^9)$ presented with bars describing objective function values, and, on the *right*, all 10 solutions generated. On the *lower-right*, best candidates, that is, solutions used in classifications and generating intermediate solutions

Fig. 3). Compared to the trial and error method or methods demanding a large database of Pareto optimal solutions, our approach makes treatment planning times shorter, and a good trade-offs between the objectives can be found to improve the treatment plan's quality. A good example of this trade-off information can be seen in Table 1: when comparing solutions $\mathbf{f}(t^1)$ and $\mathbf{f}(t^2)$, the dose in the rectum could be decreased without losing the target coverage, for example.

For comparison, in Figs. 1 (right) and 2, a treatment plan obtained with the BrachyVision® optimization tool is presented. Further, in Table 2, there are target's 90% dose value and OARs's 2 cm^3 dose volume histogram point values presented. When comparing the two solutions, it can be seen from Fig. 2 that dose volume histogram values start to decrease faster in IND-NIMBUS solution than in BrachyVision® solution. However, from Table 2 we can see that dose (2 cm^3) in bladder and rectum is smaller in solution obtained with BrachyVision®, but, at the same time, the PTV (90%) is covered with radiation better and the dose in sigma (2 cm^3) is smaller in the solution obtained with interactive multiobjective optimization. These results are one evidence more showing that the radiotherapy objectives are in conflict, and tools capable to help the decision maker (treatment planner) in navigating among different optimal treatment plans are needed.

Table 2 Comparison of target 90% values and OAR 2 cm^3 values

	Interactive method (Gy)	BrachyVision® (Gy)
Target 90%	7.32	7.23
Bladder 2 cm^3	5.24	5.06
Rectum 2 cm^3	5.64	5.01
Sigma 2 cm^3	1.38	1.61

4 Conclusions

In this paper, we have presented a new interactive multiobjective optimization approach for anatomy based 3D HDR brachytherapy optimization. In this research, the multiobjective nature of the problem has been genuinely taken into account in the problem formulation and in the interactive solution process which was directed by a treatment planner. We have demonstrated the advantages of our interactive approach by an example of a clinical gynecologic cancer.

In this study, our interactive approach has been used to determine the dwell time values needed to fulfill the prescribed dose to the tumor and to minimize dose in each organ at risk. In our approach, the decision maker's (i.e. treatment planner's) knowledge and preferences are used during the iterative optimization process to direct the search in order to find the most preferred treatment plan. This can make treatment planning times shorter and improve the treatment plan's quality. In addition, let us point out that our interactive approach is capable of handling multiple and strongly conflicting objectives in a convenient way, and thus, it offers a possibility to navigate among the obtained Pareto optimal solutions (i.e. different treatment plans).

In the presented example, there were 17 continuous decision variables and 508 sampling points. The amount of variables can easily be increased even to hundreds in more complex cases and there can be thousands of sampling points. In addition, there were only box constraints for variables. However, it is easy to add any other constraints to our interactive multiobjective optimization approach if needed. In addition, the idea of classifying objective functions is practical and computation is fast also with different numbers of objective functions. The number of objective functions can be increased, but naturally the cognitive load of the decision maker increases.

Finally, let as add that this approach brings the planning process near to the real clinical issues: with this system, the treatment planner plays directly with the compromises between a target coverage and protection of organs at risk instead of with dwell positions, dwell times, and objective weights. Whenever a trial-and-error method is used, there are no guarantees for the (Pareto) *optimality* of the final solution. Opposite to this, our approach avoids this shortcoming. These kinds of tools are not intended to replace human treatment planners, but to support them in their work.

References

Craft D, Bortfeld T (2008) How many plans are needed in an IMRT multi-objective plan database? Phys Med Biol 53:2785–2796
Craft D, Halabi T, Bortfeld T (2005) Exploration of tradeoffs in intensity-modulated radiotherapy. Phys Med Biol 50:5857–5868
Craft D, Halabi T, Shih HA, Bortfeld T (2007) An approach for practical multiobjective IMRT treatment planning. Int J Radiat Oncol Biol Phys 69:1600–1607
Ehrgott M, Winz I (2008) Interactive decision support in radiotherapy treatment planning. OR Spectrum 30:311–329
Hoffmann AL, Siem AYD, den Hertog D, Kaanders JHAM, Huizenga H (2006) Derivate-free generation and interpolation of convex Pareto optimal IMRT plans. Phys Med Biol 51:6349–6369
Holder A (2006) Partitioning multiple objective optimal solutions with applications in radiotherapy design. Optim Eng 7:501–526
Lahanas M, Baltas D (2003) Are dose calculations during dose optimization in brachytherapy necessary? Med Phys 30(9):2368–2375
Lahanas M, Baltas D, Zamboglou N (1999) Anatomy-based three-dimensional dose optimization in brachytherapy using multiobjective genetic algorithms. Med Phys 26:1904–1918
Lahanas M, Milickovic N, Papagiannopoulou M, Baltas D, Zamboglou N, Karouzakis K (2001) Application of a hybrid version of NSGA-II for multiobjective dose optimization in brachytherapy. In: Giannakoglou KC, Tsahalis DT, Périaux J, Papailiou KD, Fogarty T (eds) Evolutionary methods for design, optimization and control with applications to industrial problems. International Center for Numerical Methods in Engineering (CIMNE), Athens, Greece, pp 299–304
Lahanas M, Baltas D, Giannouli S (2003a) Global convergence analysis of fast multiobjective gradient based dose optimization algorithms for high dose rate brachytherapy. Phys Med Biol 48:599–617
Lahanas M, Baltas D, Zamboglou N (2003b) A hybrid evolutionary algorithm for multiobjective anatomy based dose optimization in HDR brachytherapy. Phys Med Biol 48:399–415
Larichev O (1992) Cognitive validity in design of decision aiding techniques. J Multicriteria Decis Anal 1:127–138
Lessard E, Pouliot J (2001) Inverse planning anatomy-based dose optimization for HDR-brachytherapy of the prostate using fast simulated annealing algorithm and dedicated objective function. Med Phys 28:773–779
Lessard E, Hsu I, Pouliot J (2002) Inverse planning for interstitial gynecologic template brachytherapy: truly anatomy-based planning. Int J Rad Oncol Biol Phys 54:1243–1251
Lessard E, Hsu I, Aubry J, Pouliot J (2006) SU-FF-T-337: multiobjective inverse planning optimization: adjustment of dose homogeneity and urethra protection in HDR-brachytherapy of the prostate. Med Phys 33:2124
Miettinen K (1999) Nonlinear multiobjective optimization. Kluwer, Boston
Miettinen K (2006) IND-NIMBUS for demanding interactive multiobjective optimization. In: Trzaskalik T (ed) Multiple criteria decision making '05. The Karol Adamiecki University of Economics, Katowice, pp 137–150
Miettinen K, Mäkelä MM (1995) Interactive bundle-based method for nondifferentiable multiobjective optimization: NIMBUS. Optimization 34:231–246
Miettinen K, Mäkelä MM (2002) On scalarizing functions in multiobjective optimization. OR Spectrum 24:193–213
Miettinen K, Mäkelä MM (2006) Synchronous approach in interactive multiobjective optimization. Eur J Oper Res 170:909–922
Milickovic N, Lahanas M, Papagiannopoulou M, Baltas D, Zamboglou N, Karouzakis K (2001) Application of multiobjective genetic algorithms in anatomy based dose optimization in brachytherapy and its comparison with deterministic algorithms. In: Giannakoglou KC, Tsahalis DT, Périaux J, Papailiou KD, Fogarty T (eds) Evolutionary methods for design,

optimization and control with applications to industrial problems. International Center for Numerical Methods in Engineering (CIMNE), Athens, Greece, pp 293–298

Milickovic N, Lahanas M, Papagiannopoulou M, Zamboglou N, Baltas D (2002) Multiobjective anatomy-based dose optimization for HDR-brachytherapy with constraint free deterministic algorithms. Phys Med Biol 47:2263–2280

Monz M, Küfer KH, Bortfeld TR, Thieke C (2008) Pareto navigation – algorithmic foundation of interactive multi-criteria IMRT planning. Phys Med Biol 53:985–998

Nath R, Anderson LL, Luxton G, Weaver KA, Williamson JF, Meigooni AS (1995) Dosimetry of interstitial brachytherapy sources: recommentations ot the AAPM radiation therapy committee task group no. 43. Med Phys 22:209–234

Rivard MJ, Coursey BM, DeWerd LA, Hanson WF, Saiful Huq M, Ibbott GS, Mitch MG, Nath R, Williamson JF (2004) Update of AAPM task group no. 43 report: a revised AAPM protocol for brachytherapy dose calculations. Med Phys 31:633–674

Romeijn HE, Dempsey JF, Li JG (2004) A unifying framework for multi-criteria fluence map optimization models. Phys Med Biol 49:1991–2013

Ruotsalainen H, Boman E, Miettinen K, Hämäläinen J (2006) Interactive multiobjective optimization for IMRT. Working Papers W-409. http://hsepubl.lib.hse.fi/pdf/wp/w409.pdf

Thieke C, Küfer KH, Monz M, Scherrer A, Alonso F, Oelfke U, Huber PE, Debus J, Bortfeld T (2007) A new concept for interactive radiotherapy planning with multicriteria optimization: first clinical evaluation. Radiother Oncol 85:292–298

Yu Y (1997) Multiobjective decision theory for computational optimization in radiation therapy. Med Phys 24:1445–1454

Yu Y, Zhang JB, Cheng G, Schell MC, Okunieff P (2000) Multi-objective optimization in radiotherapy: applications to stereotactic radiosurgery and prostate brachytherapy. Artif Intell Med 19:39–51

Multicriteria Ranking Using Weights Which Minimize the Score Range

Chris Tofallis

Abstract Various schemes have been proposed for generating a set of non-subjective weights when aggregating multiple criteria for the purposes of ranking or selecting alternatives. The maximin approach chooses the weights which maximise the lowest score (assuming there is an upper bound to scores). This is equivalent to finding the weights which minimize the maximum deviation, or range, between the worst and best scores (minimax). At first glance this seems to be an equitable way of apportioning weight, and the Rawlsian theory of justice has been cited in support.

We draw a distinction between using the maximin rule for the purpose of assessing performance, and using it for allocating resources amongst the alternatives. We show that it has a number of drawbacks which make it inappropriate for the assessment of performance. Specifically, it is tantamount to allowing the worst performers to decide the worth of the criteria so as to maximise their overall score. Furthermore, when making a selection from a list of alternatives, the final choice is highly sensitive to the removal or inclusion of alternatives whose performance is so poor that they are clearly irrelevant to the choice at hand.

1 Introduction

One of the most influential works in the area of moral and political philosophy in the last 50 years has been John Rawls's *A Theory of Justice* (1971). Rawls rejects the utilitarian idea of "the greatest good for the greatest number". This is a concept which the multi-criteria decision community would recognize as being fraught with difficulties. These include the fact that "the good" is likely to be a multi-factor concept, and that we are also dealing with multiple stakeholders holding different views. It is important to note that even if there were agreement on how to measure

C. Tofallis
The Business School, University of Hertfordshire, College Lane, Hatfield, Hertfordshire AL10 9AB, UK
e-mail: c.tofallis@herts.ac.uk

D. Jones et al. (eds.), *New Developments in Multiple Objective and Goal Programming*, Lecture Notes in Economics and Mathematical Systems 638,
DOI 10.1007/978-3-642-10354-4_9,

and then aggregate the overall good of the population, it does not follow that maximizing it would provide any form of social justice unless of course such justice was built into the definition of "the good". Rawls viewed "justice as fairness" and felt that the worst off should not be made even worse. In particular, if public resources are to be distributed unequally, then the worst off should benefit the most. Rawls referred to this as the "difference principle".

Rawls has been cited in support of using the maximin rule for weighting criteria by Pettypool and Karathanos (2004). They proposed the rule for the purpose of appraising the work of employees under a number of criteria. Butler and Williams (2002) use the maximin rule in sharing out the fixed costs associated with shared facilities. In support of it they cite work based on experiment and survey:

> A variety of fairness criteria are discussed in the seminal paper of Yaari and Bar-Hillel (1984). They conducted a series of experiments to see which of nine possible criteria were considered most fair by a sample of people questioned. In relation to needs, an allocation based on minimizing the maximum inequality was overwhelmingly considered the most fair.

One field where the minimax concept is widely used is in location problems. When choosing locations for emergency facilities (police, ambulance, firefighting) or other public offices or services, this method selects locations so as to minimize the maximum travel time or distance to any person who is being served. The method has been criticized (e.g. Ogryczak 1997) because if there is a single recipient (or a small cluster) that is located far from the vast majority, then a location may be selected which is far from all recipients. There is thus seen to be a disproportionate effect on the decision by a tiny minority of the recipients. We shall see that a similar difficulty arises when applying the minimax concept to multicriteria weighting.

The minimax objective is also used as an alternative to least squares in regression. It involves minimizing the largest deviation or residual. This is an appropriate objective if the error distribution is uniform; this can arise when the errors arise as a result of rounding, e.g. a digital measurement device will have a limited number of digits to display. This type of regression is not appropriate if there are outliers in the data, as these will severely distort the resulting model.

Another application of the maximin objective is in the allocation of highway patrol officers to districts so as to ensure that all districts experience a reduction in speeding; the aim was to maximize the minimum reduction in the number of speeding offences (Rardin 1998, p. 158). In the field of scheduling jobs numerous objectives are used, one of these is to minimize the maximum lateness (Rardin 1998, p. 605). It is also used to minimize maximum congestion or bottlenecks. Du (1996) surveys the field of minimax applications.

2 Geometric Representation

The maximin concept has been used in the assessment of performance by a number of authors. For example, Karsak and Ahiska (2005) and Karsak (2004) consider the problem of attaching weights to the various outputs (criteria of the type "more is better") when there is a single input. To create an efficiency score each output is divided by the input and then weights are attached to each of these ratios. In DEA (data envelopment analysis) each alternative has its own weights. These are chosen so as to optimize the score for that alternative. Because this method attaches different weights for each alternative, this leads to the generation of an efficient frontier which is made up of piecewise linear segments. In DEA all the alternatives on the frontier are given the same score of 100%. In an effort to increase the discrimination between such units and identify a preferred alternative, they seek a common set of weights to be used across all alternatives. These non-negative weights are chosen so as to maximize the minimum score (maximin), subject to the condition that all scores do not exceed 100%. In criterion space a set of common weights corresponds to a line or plane.

Figure 1 shows an example involving two criteria. According to DEA, points A, B and C are ranked first with the maximum score, and ABC delineates the DEA frontier. In DEA alternative P has a score given by the ratio OP/OP', where P' is the point where the ray OP intersects the frontier. Because P' lies between A and B, the corresponding weights are determined by the slope of the line AB. Point T however would be assessed relative to the line segment BC, which corresponds to a different set of criteria weights. Of the points shown in Fig. 1, P would have the lowest score. If we now depart from the piecewise frontier in favour of a single set of common weights based on the maximin rule, we shall have a single extended line frontier. We shall have to choose weights which maximize P's score, and so the frontier will be AB (extended). Notice that the particular line segment and hence weights, are chosen by reference to the worst performing alternative. This in itself is strange because the frontier is supposed to represent best practice, and yet its location is crucially influenced by an alternative displaying worst practice.

Troutt et al. (1993) use the maximin rule as a way of further ranking those alternatives which have all been given the same 100% efficiency score from a data envelopment analysis. This differs from the above in that only efficient alternatives are considered at this second stage. Hence the worst performers cannot influence

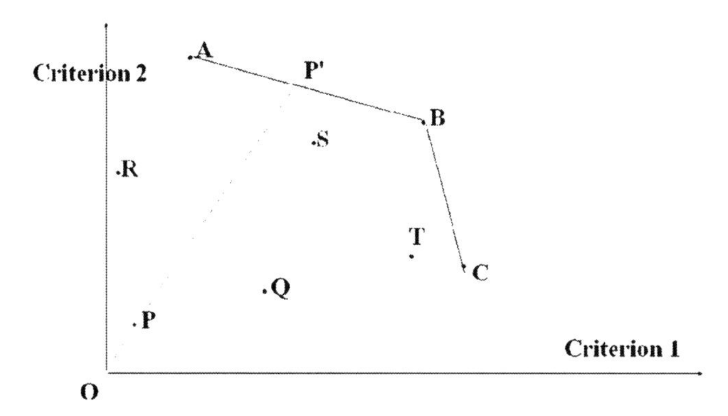

Fig. 1 Having a common set of weights (with an upper limit to the overall score) means that a line such as AB or BC acts as the frontier. The slope of such a line determines the weights

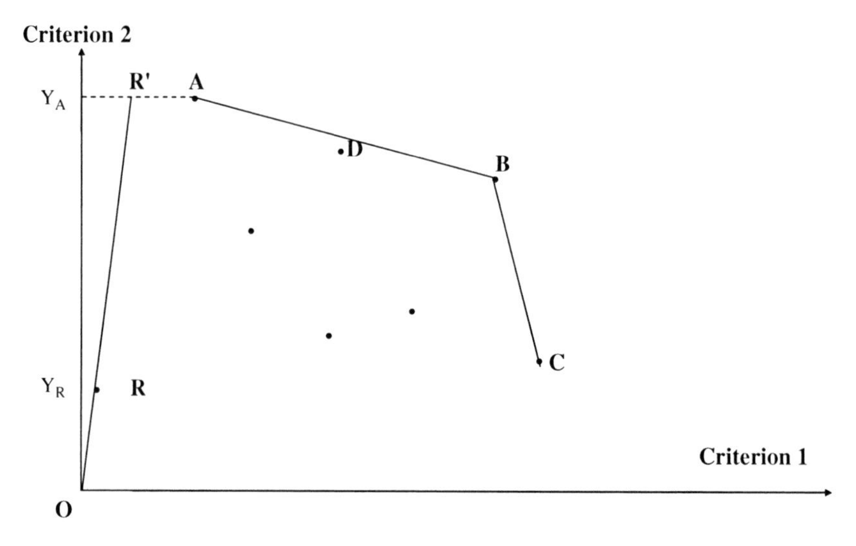

Fig. 2 When we attempt to project alternative R onto the frontier we find that its "target" (R′) does not lie between observed efficient units – i.e. it is not naturally enveloped. This leads to a horizontal frontier and a zero weight for criterion 1

the resulting weights. This is a definite improvement. The alternatives which will now influence the position of the linear frontier will be those that are at the ends of the frontier. In a two dimensional setting these will be points A and C, but in higher dimensions they will be the points on the perimeter of the frontier. Such points have very high scores in one criterion but are weak in the others, and are sometimes referred to as "mavericks". They contrast with good all-rounders. One might also include in this second stage those alternatives which are Pareto-optimal even though they do not appear on the convex hull, for example point D in Fig. 2. Such points are also "good all-rounders".

Now consider what happens when alternative P is removed from Fig. 1. Q now has the lowest score. This forces facet BC (extended) to act as the new frontier. Unit A was previously ranked first equal (maximum score), but now it slides down the rankings below B, C, T, S and Q! Karsak and Ahiska (2005) used the maximin method in a selection problem: to choose a particular piece of equipment from a number of competing alternatives. Expressed in these terms the removal of a point such as P corresponds to removing an irrelevant alternative – one that would never be selected because of its poor performance. Yet its removal causes huge changes in the rankings. This violates the axiom of decision theory known as Sen's property alpha (Sen 1969), also known as the Chernoff condition (Chernoff 1954), which states that the removal or addition of an irrelevant alternative should not affect the decision. The selection decision should be independent of irrelevant alternatives. The removal of such unwanted points could for example arise in an initial screening stage, where alternatives which do not measure up to certain minimum standards are removed from further consideration. They could also be removed from simple dominance arguments. A memorable illustration of the principle is an anecdote attributed to the philosopher Sidney Morgenbesser:

> After finishing dinner, Sidney Morgenbesser decides to order dessert. The waitress tells him he has two choices: apple pie and blueberry pie. Sidney orders the apple pie. After a few minutes the waitress returns and says that they also have cherry pie, at which point Morgenbesser says "In that case I'll have the blueberry pie."

Troutt (1997, and references therein) has written a number of papers applying the maximin approach to DEA with both multiple inputs and multiple outputs. He calls the resulting scores the MER – the maximin efficiency ratios. He makes the following observation:

> When the MER model was first discussed (without subsequent benefit of theoretical justification), some critics argued that "optimal" multipliers should not be based on least efficient units. While that criticism has intuitive merit, it may be noted that a reverse perspective is actually more fruitful. Namely, the minimum efficiency, as well as the average (or any other summary statistic) depends on the weights. Such weights or multipliers may, or may not, in general, maximize the likelihood of the resulting aggregate measure. Thus, from the maximum likelihood perspective the procedure appears intuitive. However, this apparent "contradiction of intuitions" continues to be interesting and not yet fully resolved.

Troutt and Zhang (1993) also note that "a possible objection is that the resulting weights may be overly influenced by the worst performers". They try to address this by saying "choices of weights which increase the minimum ratio frequently increase the average ratio as well, and conversely. Hence the maximin aggregation principle appears similar in expected performance to maximization of the average, which clearly depends on the performance data of the whole set of [alternatives]". This is not a persuasive argument because in the maximization of the average each point has equal influence, whereas in the maximin case this is far from being true. They also try to address the issue by first noting that using maximin leads to all scores being squeezed into the narrowest range – which is true. It is then argued that the range is a measure of dispersion, as is the variance, so one would expect similar performance to minimizing the variance of the scores, and variance does depend on all of the data. Once again, this conclusion does not follow because the calculation of variance is based on all observations whereas the range is not.

To help us understand why we would not expect similar scoring performance let us draw some parallels with methods of fitting models to data. Consider the deviations from the 100% score as being residuals, and consider that we are fitting a linear model which is constrained not to have any data points lying above it. It now becomes clear that the maximin approach corresponds to fitting using the Chebyshev or L_∞ norm, and the minimization of the average residual corresponds to the L_1 norm. It is well established that these fitting approaches produce very different models and so we cannot expect to obtain similar performance as claimed above. Specifically, the L_1 norm is less sensitive to outliers than least squares regression, whereas the Chebyshev norm is more sensitive to outliers than least squares.

3 Can the Maximin Approach Produce a Single Winning Alternative?

Pettypool and Karathanos (2004) propose the maximin approach for reward systems where there are multiple measures of reward and contribution involved. They provide a numerical example which includes three reward measures (outputs) and two contribution measures (inputs). Despite the fact that there are only seven alternatives, maximin still does not produce a single winner. Looking at Fig. 1 would seem to indicate that in the case of two outputs there will normally be three alternatives which will appear at the extremes of the score range. This is because the frontier line needs to come in as close as possible to the data points in order to keep the score range narrow. In this case P gets the lowest score, with A and B getting the highest score. As the number of criteria are increased, the higher dimensionality of the problem means that the frontier will have more dimensions and so more observations will lie upon it. Hence, although having a single set of common criteria weights will reduce the number scoring 100%, we cannot rely on the maximin approach to produce a single winner.

4 Criteria Can be Completely Ignored

Consider the set of alternatives displayed in Fig. 2. In this case R will have the lowest score as it has the worst performance on both criteria. Its score will be maximised by referring to the horizontal dashed line as a frontier. R is not fully enveloped by a pair of frontier units in the way that P was in Fig. 1, and this causes difficulties. We shall now show that using the extension of this horizontal line as a frontier to assess all other alternatives leads to criterion 1 being completely ignored in the assessment i.e. a zero weight will be applied. The demonstration involves the similar right-angled triangles $R'Y_A O$, and $RY_R O$. The angle subtended at the origin is the same for both triangles, and the cosine of this angle equates to $OY_R/OR = OY_A/OR'$. Therefore $OR/OR' = OY_R/OY_A$. But OR/OR' is precisely the score for R and OY_R/OY_A is the ratio of values on criterion 2. Thus the values on criterion 1 play no part in the assessment of R. The same argument applies to the assessment of the other alternatives.

5 Conclusion

At first sight using the maximin rule to choose a set of common weights might seem an attractive approach to an analyst. One reason is that it is not subjective, but more importantly, it reduces the likelihood of being confronted by those who fare badly from the resulting rankings – this is because the method focuses on raising their score. Thus the analyst may be able to avoid having to argue with low scorers about the weights chosen.

However, this paper has shown that a number of serious drawbacks arise when using this rule in assessing performance. Any choice of weights corresponds to deciding how much each criterion is worth in terms of utility or value. It is clear that the maximin rule is allowing those who performed worst to effectively determine these utility values. This is as sensible as allowing the worst performing student to decide how much weight to attach to each of the various assessments taken by the class.

Next consider the problem of selecting from a set of alternatives. To ease the decision a common way to reduce the number of alternatives is to use screening or filtering. This is simply the removal of those alternatives which are clearly inadequate because they do not meet certain minimal standards. This step is carried out for convenience and should not affect the final decision. However, when used in conjunction with the maximin rule such a process will remove the worst performers and so lead to a different set of weights and a different ranking of the remaining alternatives. Decisions based on the maximin rule are highly sensitive to the inclusion or exclusion of alternatives whose performance is so poor as to be completely irrelevant to the selection decision.

We also showed that when the worst performing alternative is not naturally enveloped by units on the frontier (a common occurrence with real data), then certain criteria will be given zero weight and so be completely ignored in the analysis. Given that the criteria will have been carefully selected as being appropriate at the start, it is strange that they are now being dismissed.

Whilst, the maximin approach has been used in the allocation of resources in order to reduce inequality, its use to assess such a situation of need is a different matter entirely. The stage of evaluation to determine who is most in need or most deserving is separate from the stage of assigning resources or rewards. Rawls' difference principle may be of use in the allocation stage but not in the assessment stage. To persist in using it for both would be to minimise the apparent need of the worst off and thereby reduce the resources allocated to them.

References

Butler M, Williams HP (2002) Fairness versus efficiency in charging for the use of common facilities. J Opl Res Soc 53:1324–1329

Chernoff H (1954) Rational selection of decision functions. Econometrica 22:423–443

Du Ding-Zhu (1996) Minimax and its applications. In: Horst R, Pardalos PM (eds) Handbook of global optimization. Kluwer, Dordrecht, pp 339–367

Karsak EE (2004) A practical common weight MCDM approach for emerging market selection, Proceedings of the 34th International Conference on Computers and Industrial Engineering (ICC&IE), San Francisco, California, USA, pp 159–164

Karsak EE, Ahiska SS (2005) Practical common weight multi-criteria decision-making approach with an improved discriminating power for technology selection.Int J Prod Res 43(8):1537–1554

Ogryczak W (1997) On the lexicographic minimax approach to location problems. EJOR 100: 566–585

Pettypool MD, Karathanos P (2004) An equity check. EJOR 157:465–470
Rardin RL (1998) Optimization in operations research. Prentice Hall, New Jersey
Rawls JA (1999) A theory of justice (Revised edition). Harvard University Press, Cambridge, MA
Sen AK (1969) Quasi-transitivity, rational choice and collective decisions. Rev Econ Stud 36:381–393
Troutt MD (1997) Derivation of the maximin efficiency ratio model from the maximum decisional efficiency principle. Ann Ops Res 73:323–338
Troutt MD, Zhang A (1993) The maximin efficiency ratio as a one step heuristic ranking device – comparison to two modified DEA analyses. Working paper, Dept of Management, Southern Illinois University, Carbondale, Illinois
Troutt MD, Zhang A, Pettypool MD (1993) A maximin efficiency ratio model for consensus weights and extended ranking of technically efficient units. Working paper, Dept of Management, Southern Illinois University, Carbondale, Illinois
Yaari ME, Bar-Hillel M (1984) On dividing justly. Soc Choice Welfare 1:1–24

In Search of a European Paper Industry Ranking in Terms of Sustainability by Using Binary Goal Programming

Roberto Voces, Luis Diaz-Balteiro, and Carlos Romero

Abstract Sustainability is a multidimensional concept in continuous evolution. However, the suitability of using several indicators of a diverse nature to characterise and quantify this concept has been widely accepted. Within this orientation, in this work, the paper industry's sustainability in a significant number of European countries has been analysed. To achieve this purpose, a set of economic, environmental and social indicators have been defined for the year 2004. With the help of a binary goal programming model, these indicators were aggregated into a synthetic index that measures the overall sustainability of the industry analysed. In this way, a "ranking" according to the sustainability of the paper industry in the European countries studied has been obtained.

1 Introduction

The term "sustainability" is easy to understand intuitively, although it is not at all easy to conceptualise, to measure or to formalize rigorously. Different international forums related to sustainable development have recognized that the term implies ecological and economic dimensions (Diaz-Balteiro and Romero 2008). However, from an entrepreneurial perspective, the concept of sustainability is more questionable. In fact, from a business undertaking point of view, sustainability on many occasions is linked to components related to competitiveness, innovation and the marketing of companies, and with this combination of ideas, a certain company is able to differ from its competitors in order to improve its economic performance. Thus, nowadays, the diverse environmental components of some firms are not only included through several environmental quality systems, but also in their own strategies (Aulí 2002).

R. Voces (✉)
Research Group "Economics for a Sustainable Environment", Technical University of Madrid, Madrid, Spain
e-mail: roberto.voces@upm.es

D. Jones et al. (eds.), *New Developments in Multiple Objective and Goal Programming*, Lecture Notes in Economics and Mathematical Systems 638,
DOI 10.1007/978-3-642-10354-4_10,

In this paper, we have attempted to characterize the sustainability of the paper industry at a European level, but not by trying to distinguish the firms that show tangible results in some aspects like the "triple bottom", eco-efficiency, or the installation of certain environmental management systems. On the contrary, we have defined a set of indicators that permit the characterization of the managerial reality of these industries under sustainability terms. The proposed approach has been applied to the paper industries of a significant number of European countries. To undertake this task, the methodology used has been based on a goal programming (GP) model with binary variables. This approach has been successfully applied in forestry systems (Diaz-Balteiro and Romero 2004a, b). It should be noted that this type of orientation, defining sustainability by using a set of indicators, appeared in the mid 1980s and was consolidated after the 1992 United Nations Conference on Environment and Development in Río de Janeiro. After that Conference, different lists of sustainability indicators have been proposed for their application, for instance, to different forest contexts (Castañeda 2000). However, the proposed indicators have not been defined at an entrepreneurial level. Consequently, there are few papers explicitly dealing with this topic in the forest industry. One exception to this trend is the work of Hart et al. (2000), in which different cases corresponding to multinational firms were analysed. They mainly focused on qualitative aspects, related to how some of these firms managed their forests. A similar approach can be found in Johnson and Walck (2004), who described five criteria necessary for integrating sustainability into forest industries. The complexity of selecting a representative set of key indicators has already been approached by several authors in their research on sustainability associated with forest management problems (Mendoza and Prabhu 2000a, b).

2 Sustainability Indicators

In order to define the sustainability of an industry or of a group of industries, it is necessary to measure different types of indicators: economic, social, environmental, etc. Nowadays, it is necessary to link sustainability at the entrepreneurship level not only to the existence of the firm as a simple supplier of goods with a market value, but also to another group of attributes (social, environmental) that can provide it with a higher added value as a function of the consumers' perceptions. In the last few years, these intangible attributes have been integrated into expressions like "corporate social responsibility".

Although we have incorporated all these attributes into this study, the industrial nature of the activities considered imposes the prevalence of economic indicators. Also, the scant level of the disaggregation of environment information, which still awaits an adequate treatment, should be underlined. In short, fourteen indicators encompassed in the above perspectives have been selected and are shown in Table 1. In this way, we aimed to include the different aspects of the value chain of the European paper industry which determine a greater or lesser sustainability.

Table 1 Indicators used in this study

	Indicator	Sources	Type
1	Dependence on industrial roundwood	UNECE	More is better
2	Investment rate	Eurostat	More is better
3	Intensity in labour force	Eurostat	More is better
4	Unitary average wage	Eurostat	More is better
5	Gross value added per employee	Eurostat	More is better
6	Energy efficiency	Eurostat	Less is better
7	Innovative enterprises	Eurostat; Statistik Austria	More is better
8	Effects of innovation	Eurostat; Statistik Austria	More is better
9	Acquisition of built-in technology	Eurostat	More is better
10	Patent applications	Eurostat	More is better
11	Gross value added	Eurostat	More is better
12	External competitiveness	UN Comtrade database	More is better
13	Total waste	Eurostat	Less is better
14	Environmental protection expenditure	Eurostat; Statistics Sweden; Czech Statistical Office	More is better

The selection of these indicators was conditioned, firstly, by the information available at a European level. The statistical sources used, such as Eurostat databases, are mainly of an international nature. Similarly, United Nations statistical data of wood products and international trade have been used because the paper industries are integrated into these databases. Nevertheless, when necessary, different National Offices of Statistical data have been consulted.

Next, we have analyzed the meaning of the fourteen indicators selected, which can be classified into two classes or categories: "less is better", or "more is better", since a reduction or an increment in the indicators' values supports the sustainability of the industry. The dependence of industrial roundwood gives valuable information about the different national market strategies for this input, and it is defined by the quotient between imports and apparent consumption. It should be remembered that the latter is equal to the sum of national production plus the imports less the exports.

The investment rate provides information on the intensity in the use of the capital factor for this industry in each country, measured as the quotient between investment and value added at factor cost. On the other hand, the following indicators present, direct or indirectly, labour use as a production factor. Thus, the intensity of the labour force (percentage of labour costs in total production) gives information on the intensity in the use of labour as a production factor for the paper industry in each country. The more traditional sectors, of a lesser complexity and vitality, also use this factor more intensively (Fonfría 2004). For that reason, in this study it was preferable for this indicator to reach its lowest possible value. Conversely, the unitary average wage indicator shows workers' earnings for this industrial sector in each country. Without analysing the differences associated with the national income per capita, a higher value of this indicator is considered as being more sustainable from a social point of view. Finally, the gross added value per person employed shows an approach to the traditional "average product of labour" concept.

Regarding energy efficiency, this indicator represents a marginal cost, because it covers the amount of energy that it is necessary to buy in order to obtain an additional metric ton of product. Logically, a greater sustainability is reached when the value of this indicator is a low one.

Next, we show four indicators related to innovation. First, the percentage of innovative firms with regard to the total number of firms could be the indicator that shows the penetration rate of innovative activities in the paper industry. Also, the percentage of the total turnover of the paper industry in each country due to innovative firms supplies information about the real importance of those innovative activities in the final outputs of this sector. Actually, the number of patent applications to the European Patent Office in the reference year (2003) is a widely used indicator of the output due to the innovative activities developed in each country, and it has been used in this research. Finally, it has been considered to be appropriate to incorporate the acquisition of built-in technology into this group of indicators, because this is the principal way to incorporate innovation, mainly in small and medium-sized firms. These indicators have been considered as belonging explicitly to the category "more is better", since the higher the figures, the more the paper industry will be sustainable. This is because it is usually recognized that a good way to achieve a greater sustainability of firms could be by increasing the results associated with the I+D+i (Paech 2007).

The gross value added as a percentage with respect to the paper industry in the manufacturing sector constitutes an indicator that shows the relative weight of this industrial sector in the total manufacturing activity of each country. It has been considered that a reduced contribution of value added implies a reduced allocation of resources compared to other more productive and dynamic industrial sectors. In this context, a complementary indicator could be the revealed comparative advantage index (Balassa index). This has been defined as the relationship between the importance of the exports of a certain industrial sector with respect to the total industrial exports in a particular country, and, over a wider area that might be the whole world, Europe, or, in this case, the cluster of European countries analyzed. It represents an external competitiveness indicator, and if this index has a larger value than the unit, a competitive advantage does exist, or, in a contrary sense, it does not.

Finally, in this investigation we included two indicators related to some environmental characteristics of these firms. First, the waste generated by them gives information on the pollutants produced by their industrial activity. To allow a comparison between the different countries, this figure is divided up between the value added corresponding to each specific paper industry. It has been assumed that "less is better", because, in this way, the sustainability of these firms increases. The last indicator in Table 1 shows the quotient between the total current expenses for environmental protection and the number of employees. Here, only the expenditure on environment protection that exclusively affects the period in which it was incurred, without any future economic projection, will be included. For the purpose of comparing the different figures corresponding to the European countries included in this analysis, this value is distributed between the number of employees. It should be mentioned that, at the moment of carrying out this study, the data corresponding to Italy for the year 2004 was not available, so that, and only for this indicator and country, the year adopted was 2003.

3 Methodology

As specified above, for the purpose of aggregating the different sustainability indicators previously defined into a synthetic index that measures the sustainability of the different countries, an analytic procedure based on GP with binary variables was used (Diaz-Balteiro and Romero 2004b). Thus, we have considered the general case in which there are n countries, evaluating each one of them according to m sustainability indicators, applying the analysis made in the previous section. In this context, a key question was to determine the ranking of the n countries in terms of sustainability.

On these lines, the sustainability indicators were measured in different units, and also with very different absolute values. For that reason, a first stage in our work consisted of appropriately normalising the m indicators. We did so by applying the procedure suggested in Diaz-Balteiro and Romero (2004a, b). The proposed procedure adapted to our context is summed up in the following formulae:

$$\overline{R}_{ij} = 1 - \frac{R_j^* - R_{ij}}{R_j^* - R_{*j}} = \frac{R_{ij} - R_{*j}}{R_j^* - R_{*j}} \quad \forall i, j \tag{1}$$

Where $\overline{R}_{ij}$ is the normalised value reached by the ith country when it is evaluated according to the jth indicator; R_{ij} is the result reached by the ith country when it is evaluated according to the jth indicator; R_j^* is the optimum or ideal value for the jth sustainability indicator. This ideal value represents the maximum value if the indicator is of the "more is better" type, or the minimum value if the indicator is of the "less is better" type. In the same way, R_{*j} is the worst value or anti-ideal value for the jth sustainability indicator.

With this normalisation system, the indicators do not have any dimension and they are all them bounded between 0 and 1. The same procedure was applied in order to normalise the aspiration levels ("targets") of the different indicators. These aspiration levels are exogenous and they are determined by means of expert judgements, as well as from the experience accumulated by the authors. Once this point has been achieved, the following GP model was defined:
Achievement function:

$$Min \quad \sum_{j=1}^{m} (\alpha_j n_j + \beta_j p_j) \tag{2}$$

Goal and constraints:

$$\sum_{i=1}^{n} \overline{R_{ij}} X_i + n_j - p_j = \overline{t_j} \quad j \in \{1 \ldots m\} \tag{3}$$

$$\sum_{i=1}^{n} X_i = 1$$

$$X_i \in \{0, 1\} \quad i \in \{1, \ldots, n\} \tag{4}$$

$$\mathbf{n} \geq \mathbf{0} \quad \mathbf{p} \geq \mathbf{0}$$

where n_j y p_j are the deviation variables that measure the discrepancies between the value reached by the jth indicator with respect to the aspiration level $\overline{t_j}$. On the other hand, α_j and β_j are the preferential weights associated with both deviation variables. X_i are binary variables that take on the value 1 if the ith country is chosen, otherwise they take on the value 0. By solving the model (2)–(4) the country with the most sustainable paper industry was determined. Applying this procedure in an iterative way, the "ranking" of the countries analysed in sustainable terms was obtained.

In short, the application of the preceding model provides an apparently attractive solution, because it implies the greatest aggregated effectiveness. Nevertheless, this kind of solution can produce highly deviated results for some of the indicators selected, which could be unacceptable when classifying the sustainability for this industry in the countries chosen. To solve this problem, another GP model has been proposed in order to obtain the most balanced solution associated with the achievement of the different goals (Tamiz et al. 1998), with the following analytic expression:
Achievement function

$$Min\ D$$

Goal and constraints:

$$(\alpha_j n_j + \beta_j p_j) - D \leq 0 \tag{5}$$

where D represents the maximum deviation between an indicator and its aspiration level. However, if we wished to merge both GP models in only one single formulation, then it would be necessary to set up an extended GP (EGP) model, with the following analytic expression (Romero 2004):
Achievement function:

$$Min\ (1 - \lambda)\ D + \lambda \sum_{j=1}^{m} (\alpha_j n_j + \beta_j p_j) \tag{6}$$

subject to:
Goals and constraints from the model defined by (5).

In this case, for $\lambda = 1$, the most efficient solution, or the one with a better average result has been obtained, while for $\lambda = 0$ the most balanced solution has been elicited. For intermediate values of the control parameter λ, compromises between both solutions, if they exist, will be obtained. For the resolution of this model, the software LINGO 10 (Lindo Systems 2007) was applied.

4 Results and Conclusions

Once the national values and the normalised aspiration levels for the 14 indicators used in this analysis had been obtained, the EGP model shown in the (6) was applied. Table 2 shows the final ranking of the 17 countries, according to the different values of control parameter λ. In the first place, it can be verified how the ranking associated with the most efficient solution ($\lambda = 1$), is different to the ranking associated with the most balanced solution ($\lambda = 0$). These differences are in some cases remarkable, as can be seen in countries like Romania or the Czech Republic, which notably change their position in the ranking. The country with the most sustainable paper industry was either Portugal or Sweden, according to the different solutions obtained. Conversely, the country with the least sustainable paper industry was Latvia.

It has also been attempted to find out the sensitivity of the solution shown in Table 2, when the preferential weights conferred on some indicators were modified. For this purpose, a sensitivity analysis was developed for four indicators, while the other weights corresponding to the rest of the indicators remained unchanged. The results obtained were different depending on the indicator selected. Thus, whereas changes in the weights associated with the indicator related to expenses for

Table 2 Results according to parameter λ values

$\lambda = 0$	$\lambda = 1$
Portugal	Sweden
Romania	Portugal
Sweden	Finland
The Slovak Republic	Austria
Finland	Germany
Czech Republic	France
United Kingdom	Spain
Hungary	United Kingdom
Lithuania	Estonia
France	The Slovak Republic
Austria	Italy
Spain	Cyprus
Italy	Hungary
Cyprus	Czech Republic
Germany	Lithuania
Estonia	Romania
Latvia	Latvia

environmental protection or to the gross value added per employee did not cause any remarkable changes in the ranking, if a larger weight was given to the indicator measuring the waste generated as a function of the gross value added, the solution was modified irrespective of the value of control parameter λ.

We should like to end this paper by indicating that the procedure followed to obtain an overall measurement of paper industry sustainability in some European countries permits an easy integration of different indicators of a highly diverse nature. Thus, and remembering that, to a certain extent, the selection of those indicators has been conditioned by the data available, it would be necessary to stress that the GP method applied has shown itself to be very flexible, allowing us to obtain the best solution from an aggregated point of view, the best solution from a balanced perspective, or compromises between these two solutions. Finally, this work could be extended in several directions. For example, the models could be replicated by trying to introduce different preferential weights for each indicator considered. These weights could be obtained by means of judgements from experts. Another possible expansion of this research would consist of adapting the analysis at a more disaggregated level, for instance at a managerial one, or analysing in more detail certain industrial subgroups.

Acknowledgements This work was funded by the Spanish Ministry of Education and Science under project AGL2008-01457, and by the Autonomous Government of Madrid under project Q060705083. Thanks go to Diana Badder for editing the English.

References

Aulí E (2002) Integración de los factores ambientales en las estrategias empresariales. Boletín Económico del ICE 800:139–148

Castañeda F (2000) Criteria and indicators for sustainable forest management: international processes, current status and the way ahead. Unasylva 203:34–40

Diaz-Balteiro L, Romero C (2004a) In search of a natural systems sustainability index. Ecol Econ 49:401–405

Diaz-Balteiro L, Romero C (2004b) Sustainability of forest management plans. A discrete goal programming approach. J Environ Manag 71: 351–359

Diaz-Balteiro L, Romero C (2008) Making forestry decisions with multiple criteria: a review and an assessment. For Ecol Manag 255:3222–3241

Fonfría A (2004) La innovación tecnológica en los sectores tradicionales españoles Econ Indus 355–356:37–46

Hart S, Arnold M, Day R (2000) The business of sustainable forestry: Meshing operations with strategic purpose. Interfaces 30:234–254

Johnson A, Walck D (2004) Certifified success: integrating sustainability into corporate management systems J Forest 102:32–39

Lindo Systems (2007) LINGO v. 10.0, Chicago

Mendoza GA, Prabhu R (2000a) Development of a methodology for selecting criteria and indicators of sustainable forest management: a case study on participatory assessment. Env Manag 26:659–673

Mendoza GA, Prabhu R (2000b) Multiple criteria decision making approaches to assessing forest sustainability using criteria and indicators: a case study For Ecol Manag 131:107–126

Paech N (2007) Directional certainty in sustainability-oriented innovation management. In: Lehmann-Waffenschmidt M (ed) Innovations towards sustainability. Conditions and consequences. Physica, Heidelberg, pp 121–139

Romero C (2004) A general structure of achievement function for a goal programming model. Eur J Oper Res 153:675–686

Tamiz M, Jones DF, Romero C (1998) Goal programming for decision making: an overview of the current state-of-the-art. Eur J Oper Res 111:569–581

Nurse Scheduling by Fuzzy Goal Programming

Mehrdad Tamiz and Mohammed Ali Yaghoobi

Abstract This paper develops a fuzzy goal programming model to plan and allocate nurses to hospital wards. It is applied to a hospital ward in Kerman, Iran. The monthly schedule developed by the proposed model takes into account both hospital objectives and nurses' preferences. First, the nurses' preferences are obtained via a questionnaire. Then, the objectives and the nurses' preferences are divided into two major parts in the proposed model: hard constraints and fuzzy goals. The resulting schedule, by solving the proposed model, is implemented and is much preferred to the current schedule. Moreover, the proposed model is easily applicable, takes a short time to solve, and can be extended to include other objectives or preferences.

1 Introduction

Nurse scheduling problems are challenging problems in large hospitals and difficult to resolve fairly (Bester et al. 2007). The aim of nurse scheduling is usually to assign different kinds of working shifts to nurses having different skills in terms of some legal and policy constraints, and attempting to achieve an acceptable trade off between some objectives. Objectives are due to hospital requirements or nurses' preferences. Amongst legal and policy constraints are continuous patient care and minimum number of nurses with appropriate nursing skills in each shift, minimum or maximum restrictions on the number of working hours per month. On the other hand, some of nurses' preferences and hospital objectives are more day shifts than night shifts, more day shifts than evening shifts and at most three continuous night shifts. Nurse' preferences can be attained by distributing a questionnaire to all of them and analyzing it by some appropriate statistical tools. Because of different objectives, model of nurse scheduling usually leads to a multi-objective programming model.

M.A. Yaghoobi (✉)
Department of Statistics, Faculty of Mathematics and Computer, Shahid Bahonar University of Kerman 76169-14111, Kerman, Iran
e-mail: yaghoobi@mail.uk.ac.ir

D. Jones et al. (eds.), *New Developments in Multiple Objective and Goal Programming*, Lecture Notes in Economics and Mathematical Systems 638,
DOI 10.1007/978-3-642-10354-4_11,

Several approaches are suggested to deal with a multi-objective programming model in healthcare (Aickelin and White 2004; Azaeiz and Al-Sharif 2005; Bard and Purnomo 2005; Bester et al. 2007; Blake and Carter 2002). Blake and Carter (2002) describe the application of goal programming (GP) to strategic resource allocation in an acute care hospital. Chu et al. (2000) also used GP models for nurse allocation for maternal and child health services in Hong Kong aiming at an equitable allocation of nurses of different ranks, with different work functions, at different centers. Azaiez and Al-Sharif (2005) developed a zero-one GP model adapted to Riyadh Al-Kharj hospital program in Saudi Arabia. Topaloglu and Ozkarahan (2004) suggested an implicit GP model for the tour scheduling problem considering the employee work preferences. A general review of nurse scheduling models and solution approaches can be found in Bester et al. (2007), Burke et al. (2004), and Sitompul and Randhawa (1990).

As stated above, there are many researches that use GP for nurse scheduling. However, in conventional GP models the decision maker is required to specify a precise aspiration level for each of the objectives. In general, especially in large-scale problems, this is a difficult task for the decision maker(s). Applying fuzzy set theory in GP can help the decision maker to specify imprecise aspiration levels (Narasimhan 1980; Yaghoobi and Tamiz 2007). This paper develops a fuzzy goal programming (FGP) model for nurse scheduling. The model is based on the current requirement of Ayatolah Kashani (AK) hospital in Kerman, Iran. In the developed model legal and policy constraints as well as nurses' preferences are divided into two major parts: hard constraints and fuzzy goals. Then, a similar approach to the one developed by the authors (Yaghoobi et al. 2008; Yaghoobi and Tamiz 2007) is employed for solving the resulting model. Comparison of the schedule made by the developed model with the one made manually shows an improved performance.

2 Problem Statement: The Case Study

Our study is based on the current situation of AK. It is a large hospital involving many wards. Every ward has a certain number of nurses of different ranks or levels. It is the task of each ward to schedule its nurses' duties. In fact, a specified staff member of each ward, usually head nurse, assigns nurses to the shifts in terms of individual nurse preferences and hospital requirements.

Our focus is on the maternity ward (MW) which is active 24 h a day. However, due to hospital policy, each day is partitioned into three shifts: day shift, evening shift, and night shift. Day shift is from 6 AM to 12 PM; evening shift is from 12 to 6 PM; night shift is from 6 PM to 6 AM of the next day. A minimum number of nurses are required to cover the shifts in MW; four for day shift; three for evening shift; four for night shift. The total number of nurses that work in MW is 19. Moreover, they are divided into two levels: level A and level B. Level A includes nurses with 15 or more years of experience. On the other hand, level B includes nurses with less than 15 years of experience. Due to the acute task for nurses in MW, it is required to have at least two nurses of level A in any night shift, and at least one in each day and evening shifts.

Schedule is made for a 1 month period at a time. Prior to scheduling, each nurse is asked to select (optional) 3 days off (holidays) during the month. The schedule must assign only one shift to each nurse per day. Also, a night shift must be followed by a day off. Note that a day off can be selected by a nurse, assigned by the schedule after a night shift, or assigned by the schedule to avoid extra nurses than the required minimum for each shift. Every nurse must work for 176 h per month and any additional hours are paid overtime.

It is the task of head nurse in MW to allocate nurses to different shifts fairly so that both nurses' preferences and hospital policies are achieved by as much as possible. At present, MW nurse scheduling is performed manually. Indeed, it is done through trial and error. It takes two to three working days for a head nurse to build the schedule each month.

2.1 Legal and Policy Restrictions

The legal and policy restrictions are as follows:

- Three shifts a day must be covered.
- A nurse must not work for more than six consecutive days.
- A nurse must not be assigned to more than one shift per day.
- The minimum demand for the number of nurses must be met for each shift.
- A nurse must not work the day after a night shift.
- Some nurses cannot do night shifts due to health reasons.
- A nurse must not work for more than three consecutive night shifts.
- Nurse preferences for 3 days off selection must be taken into account.

Also there are some other policies that the managers wish to be satisfied if possible. They are as follows:

- More day and evening shifts than night shifts assign per month per nurse.
- More day shifts than evening shifts assign per month per nurse.
- Day shift should not follow evening shift in the previous day.
- All nurses get the same number of night shifts and the same number of weekends off.

2.2 Nurses' Preferences

To obtain nurses' preferences, a questionnaire was distributed to all 19 nurses. The following set of preferences was established.

- All nurses like to get the same share of night shifts.
- Almost all nurses prefer no more than three consecutive night shifts.

- All nurses prefer to get the same share of day offs during weekends and public holidays.
- Almost all nurses prefer to have more day shifts than evening shifts.
- Almost all nurses do not prefer a day shift following an evening shift.

3 Fuzzy Goal Programming

A useful tool for dealing with imprecision is fuzzy set theory (Zadeh 1965). An objective with an imprecise aspiration level can be treated as a fuzzy goal. Initially, Narasimhan incorporated fuzzy set theory in GP in 1980 and presented an FGP model (Narasimhan 1980). Hannan simplified the Narasimhan's method to an equivalent simple linear programming in 1981 (Hannan 1981). These pioneering works led to extensive research in the use and application of FGP to real life problems. To solve FGP problems various models based on different approaches have been proposed. A survey and classification of FGP models has been presented by Chanas and Kuchta (2002). There are three types of fuzzy goals which are the most common. The following FGP model contains these fuzzy goals.

$$\begin{aligned} \mathit{OPTIMIZE}\ c^i x &\underset{\sim}{\leq} g_i \quad && i = 1, \ldots, i_0 \\ c^i x &\underset{\sim}{\geq} g_i && i = i_0 + 1, \ldots, j_0 \\ c^i x &\underset{\sim}{=} g_i && i = j_0 + 1, \ldots, K \\ x &\in X, \end{aligned} \tag{1}$$

where OPTIMIZE means finding an optimal decision x such that all fuzzy goals are satisfied (Hannan 1981; Yaghoobi and Tamiz 2007), $c^i x = \sum_{j=1}^{n} c_{ij} x_j, i = 1, \ldots, K$, g_i is the aspiration level for the goal i, X is an optional set of hard constraints as found in linear programming and the symbol $\sim$ is a fuzzifier representing the imprecise fashion in which the goals are stated.

The fuzzy goals can be identified as fuzzy sets defined over the feasible set with the membership functions. For the three types of fuzzy goals in model (1), linear membership functions are defined as follows (Hannan 1981; Narasimhan 1980; Yaghoobi and Tamiz 2007):

$$\mu_i = \begin{cases} 1 & c^i x \leq g_i \\ 1 - \dfrac{c^i x - b_i}{\Delta_{iR}} & g_i \leq c^i x \leq g_i + \Delta_{iR} \\ 0 & c^i x \geq g_i + \Delta_{iR} \end{cases} \quad i = 1, \ldots, i_0, \tag{2}$$

$$\mu_i = \begin{cases} 1 & c^i x \geq g_i \\ 1 - \dfrac{b_i - c^i x}{\Delta_{iL}} & g_i - \Delta_{iL} \leq c^i x \leq g_i \\ 0 & c^i x \leq g_i - \Delta_{iL} \end{cases} \quad i = i_0 + 1, \ldots, j_0, \tag{3}$$

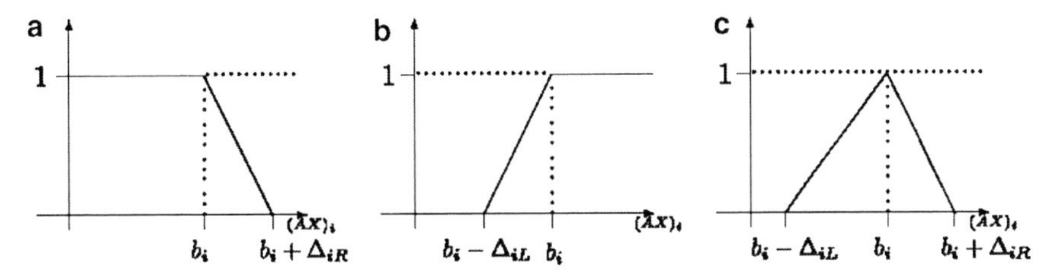

Fig. 1 Linear membership functions: (**a**) $i = 1, \ldots, i_0$, (**b**) $i = i_0 + 1, \ldots, j_0$ and (**c**) $i = j_0 + 1, \ldots, K$

$$\mu_i = \begin{cases} 0 & c^i x \le g_i - \Delta_{iL} \\ 1 - \dfrac{b_i - c^i x}{\Delta_{iL}} & g_i - \Delta_{iL} \le c^i x \le g_i \\ 1 - \dfrac{c^i x - b_i}{\Delta_{iR}} & g_i \le c^i x \le g_i + \Delta_{iR} \\ 0 & c^i x \ge g_i + \Delta_{iR} \end{cases} \quad i = j_0 + 1, \ldots, K, \tag{4}$$

where Δ_{iL} and Δ_{iR} are chosen constants of the maximum admissible violations from the aspiration level g_i. They are either subjectively chosen by the decision maker or are tolerances in a technical process. The above membership functions are depicted in Fig. 1.

Indeed, all of the above membership functions belong to the class of problems with piecewise linear concave membership functions.

Recently in (Yaghoobi et al. 2008), a model based on the weighted variant of GP (Romero 2004) is proposed to solve the FGP model (1) as follows:

$$\min\ z = \sum_{i=1}^{i_0} w_i \frac{p_i}{\Delta_{iR}} + \sum_{i=i_0+1}^{j_0} w_i \frac{n_i}{\Delta_{iL}} + \sum_{i=j_0+1}^{K} w_i \left(\frac{p_i}{\Delta_{iR}} + \frac{n_i}{\Delta_{iL}} \right)$$

s.t.

$$\begin{aligned} c^i x - p_i &\le g_i & i &= 1, \ldots, i_0 \\ \mu_i + \frac{p_i}{\Delta_{iR}} &= 1 & i &= 1, \ldots, i_0 \\ c^i x + n_i &\ge g_i & i &= i_0 + 1, \ldots, j_0 \\ \mu_i + \frac{n_i}{\Delta_{iL}} &= 1 & i &= i_0 + 1, \ldots, j_0 \\ c^i x + n_i - p_i &= g_i & i &= j_0 + 1, \ldots, K \\ \mu_i + \frac{p_i}{\Delta_{iR}} + \frac{n_i}{\Delta_{iL}} &= 1 & i &= j_0 + 1, \ldots, K \\ \mu_i, n_i, p_i &\ge 0 & &\forall i \\ x &\in X, \end{aligned} \tag{5}$$

where w_i denotes the weight of the ith fuzzy goal and μ_i is a model variable which determines the degree of membership function for the ith fuzzy goal (Yaghoobi et al. 2008). n_i and p_i are negative and positive deviational variables.

For this case study, we use the lexicographic variant of GP (Romero 2004) instead of weighted GP as used in model (5). Therefore, to solve the FGP model (1) the following achievement function is used over the constraints of model (5).

$$Lex\min\ z = \left(\sum_{i=1}^{i_0} \frac{p_i}{\Delta_{iR}},\ \sum_{i=i_0+1}^{j_0} \frac{n_i}{\Delta_{iL}},\ \sum_{i=j_0+1}^{K} \left(\frac{p_i}{\Delta_{iR}} + \frac{n_i}{\Delta_{iL}} \right) \right) \tag{6}$$

4 The Fuzzy Goal Programming Model for Scheduling Model

The aim of this section is to develop the FGP model for nurse scheduling in MW. The model takes into account both hospital objectives and nurses' preferences as well as the legal and policy restrictions. The following binary (0,1) decision variables are used in the model.

$XD_{i,j}$ = 1 if nurse j is working during day shift in day i.
= 0 otherwise.
$XE_{i,j}$ = 1 if nurse j is working during evening shift in day i.
= 0 otherwise.
$XN_{i,j}$ = 1 if nurse j is working during night shift in day i.
= 0 otherwise.
$XX_{i,j}$ = 1 if nurse j has a day off after a night shift in day i.
= 0 otherwise.
$XT_{i,j}$ = 1 if nurse j has a day off (holiday assigned by the schedule) in day i;
0 otherwise.
$XH_{i,j}$ = 1 if nurse j has a day off (holiday selected by the nurse) in day i;
0 otherwise.

Legal and policy restrictions, hospital objectives and nurses' preferences are categorized in two major groups: hard constraints and fuzzy goals. Note that in the model under consideration, there are 19 nurses ($j = 1, \ldots, 19$) and 30 days in the month ($i = 1, \ldots, 30$)

4.1 Hard Constraints

Constraint set 1: States that the number of nurses in every day shift should be equal to 4.

$$\sum_{j=1}^{19} XD_{i,j} = 4, \qquad i = 1, \ldots, 30. \tag{7}$$

Constraint set 2: States that the number of nurses in every evening shift should be equal to 3.

$$\sum_{j=1}^{19} XE_{i,j} = 3, \qquad i = 1, \ldots, 30. \tag{8}$$

Constraint set 3: States that the number of nurses in every night shift should be equal to 4.

$$\sum_{j=1}^{19} XN_{i,j} = 4, \qquad i = 1, \ldots, 30. \tag{9}$$

Constraint set 4: Since a nurse must not work the day after a night shift, every day 4 nurses should have a day off.

$$\sum_{j=1}^{19} XX_{i,j} = 4, \qquad i = 1, \ldots, 30. \tag{10}$$

Constraint set 5: Assigning one shift per day per nurse.

$$XM_{i,j} + XE_{i,j} + XN_{i,j} + XX_{i,j} + XT_{i,j} + XH_{i,j} = 1, \quad j = 1, \ldots, 19; \; i = 1, \ldots, 30. \tag{11}$$

Constraint set 6: Avoid more than six consecutive work days for every nurse.

$$\sum_{s=0}^{5} \left(XX_{i+s,j} + XT_{i+s,j} + XH_{i+s,j}\right) \geq 1, \quad j = 1, \ldots, 19; i = 1, \ldots, 25. \tag{12}$$

Constraint set 7: Every nurse is entitled to have at least 1 day off during the weekends and public holidays in the month.

$$XT_{3,j} + XT_{4,j} + XT_{5,j} + XT_{11,j} + XT_{12,j} + XT_{18,j} + XT_{19,j} + XT_{25,j} + XT_{26,j} + XT_{27,j} \geq 1, \; j = 1, \ldots, 19. \tag{13}$$

Note that the indices of the variables in the above constraint refer to the weekend and public holiday days in the calendar month of the planning month. That is, days 4, 5, 11, 12, 18, 19, 25 and 26 are weekends, and days 3 and 27 are public holidays.

Constraint set 8: Every nurse should have a day off after a night shift.

$$XN_{i,j} - XX_{i+1,j} = 0, \qquad j = 1, \ldots, 19; i = 1, \ldots, 29. \tag{14}$$

Constraint set 9: No more than three consecutive night shifts.

$$XN_{i-6,j} + XN_{i-4,j} + XN_{i-2,j} + XN_{i,j} \leq 3, \qquad j = 1, \ldots, 19; \; i = 7, \ldots, 30. \tag{15}$$

Constraint set 10: For fairness, scheduling maximum and/or minimum working shifts are considered per nurse per month as follows:

$$\sum_{i=1}^{30} XD_{i,j} \geq 3, \qquad j = 1, \ldots, 19, \tag{16}$$

$$\sum_{i=1}^{30} XE_{i,j} \geq 3, \qquad j = 1, \ldots, 19, \tag{17}$$

$$\sum_{i=1}^{30} XN_{i,j} \geq 6, \qquad j = 1, \ldots, 19;\ j \neq 15, \tag{18}$$

$$\sum_{i=1}^{30} XN_{i,j} \leq 7, \qquad j = 1, \ldots, 19, \tag{19}$$

$$\sum_{i=1}^{30} XT_{i,j} \leq 6, \qquad j = 1, \ldots, 19, \tag{20}$$

$$\sum_{i=1}^{30} XH_{i,j} \leq 3, \qquad j = 1, \ldots, 19. \tag{21}$$

It should be noted that nurse 15 cannot work night shifts due to a specific illness. The following constraint exempts nurse 15 from night shifts.

$$\sum_{i=1}^{30} XN_{i,15} = 0. \tag{22}$$

Constraint set 11: The required minimum number of nurses of level A per shift is determined by the following constraints.

$$\sum_{j=1}^{10} XD_{i,j} \geq 1 \qquad i = 1, \ldots, 30, \tag{23}$$

$$\sum_{j=1}^{10} XE_{i,j} \geq 1 \qquad i = 1, \ldots, 30, \tag{24}$$

$$\sum_{j=1}^{10} XN_{i,j} \geq 2 \qquad i = 1, \ldots, 30. \tag{25}$$

It should be noted that $\mathrm{j} = 1, \ldots, 10$ represent nurses belonging to level A and $\mathrm{j} = 11, \ldots, 19$ represent level B nurses.

Constraint set 12: Prior to scheduling every nurse can select up to 3 days off as leave. As an example, the following constraints show that nurses 3 and 6 requested day 28, 29 and 2, 3 off respectively:

$$\begin{aligned} XH_{28,3} &= 1, \\ XH_{29,3} &= 1, \\ XH_{2,6} &= 1, \\ XH_{3,6} &= 1. \end{aligned} \tag{26}$$

The other requests for leave are treated in the same way.

4.2 *Fuzzy Goals*

Fuzzy goal 1: One of the important issues explored from the questionnaire shows that the nurses are very sensitive to have a fair distribution of days off. Because of the problem structure, it is almost impossible to assign equal number of days off to all nurses. However, we can try to do that as much as possible by using a fuzzy goal as follows:

$$\sum_{i=1}^{30} \left(XT_{i,j} + XH_{i,j}\right) \underset{\sim}{=} 7 \qquad j = 1, \ldots, 19. \tag{27}$$

The above fuzzy goal states that the number of days off for every nurse should be approximately seven, where we consider a maximum of 2 days for admissible violations. Therefore, the above fuzzy goal based on model (5) should be rewritten as follows:

$$\begin{aligned} \sum_{i=1}^{30} \left(XT_{i,j} + XH_{i,j}\right) + n1_j - p1_j &= 7 \quad & j &= 1, \ldots, 19, \\ \frac{n1_j}{2} + \frac{p1_j}{2} &\le 1 \quad & j &= 1, \ldots, 19, \\ n1_j, p1_j &\ge 0 \quad & j &= 1, \ldots, 19, \end{aligned} \tag{28}$$

and $\sum_{j=1}^{19} \left(\frac{n1_j}{2} + \frac{p1_j}{2}\right)$ should be minimized in the achievement function.

Fuzzy goal 2: One of the major nurses' preferences is to have more day shifts than evening shifts. Furthermore, the number of day shifts required by the MW in a month is more than evening shifts. Thus, it is expected that a fair schedule should assign more day shifts than evening shifts per nurse. To this end, the following fuzzy goal is proposed. It should be noted that fuzzy goal (29) tries to assign more day shifts than evening shifts in an almost equally manner to every nurse.

$$\sum_{i=1}^{30} \left(XD_{i,j} - XE_{i,j}\right) \underset{\sim}{=} 2 \qquad j = 1, \ldots, 19. \tag{29}$$

In fact, fuzzy goal (29) states that every nurse should have approximately 2 day shifts more than evening shifts per month. A maximum of 1 day is considered for admissible violations. Therefore, the above fuzzy goal based on model (5) should be rewritten as follows:

$$\begin{aligned} \sum_{i=1}^{30} \left(XM_{i,j} - XE_{i,j}\right) + n2_j - p2_j &= 2 \quad j = 1,\dots,19, \\ \frac{n2_j}{1} + \frac{p2_j}{1} &\le 1 \quad j = 1,\dots,19, \\ n2_j, p2_j &\ge 0 \quad j = 1,\dots,19, \end{aligned} \tag{30}$$

and $\sum_{j=1}^{19} \left(\frac{n2_j}{1} + \frac{p2_j}{1}\right)$ should be minimized in the achievement function.

Fuzzy goal 3: Another nurses' preference is to have a very few, if any, day shifts following an evening shift. To take into account this preference, a fuzzy goal is suggested. At first, the day shifts follow an evening shift are counted using the following equation.

$$\begin{aligned} XE_{i,j} + XD_{i+1,j} + n3_{i,j} - p3_{i,j} &= 1. \quad i = 1,\dots,29;\ j = 1,\dots,19, \\ n3_{i,j}, p3_{i,j} &\ge 0 \quad j = 1,\dots,19. \end{aligned} \tag{31}$$

Indeed, when $p3_{i,j}$ equals 1 in (31), it shows that 1 day shift (in day $i+1$) following an evening shift is assigned to nurse j. Thus, $\sum_{i=1}^{29} p3_{i,j}$ counts the number of day shifts following evening shifts for nurse j. Hence, to avoid day shift following an evening shift a fuzzy goal is suggested as follows:

$$\sum_{i=1}^{29} p3_{i,j} \lesssim 0 \qquad j = 1,\dots,19. \tag{32}$$

This fuzzy goal states that the number of day shifts following evening shifts should be approximately 0. A maximum of 3 days is considered for admissible violation. Therefore, the above fuzzy goal based on model (5) should be rewritten as follows:

$$\begin{aligned} \sum_{i=1}^{30} p3_{i,j} - p4_j &= 0 \quad j = 1,\dots,19, \\ \frac{p4_j}{3} &\le 1 \quad j = 1,\dots,19, \\ p4_j &\ge 0 \quad j = 1,\dots,19, \end{aligned} \tag{33}$$

and $\sum_{j=1}^{19} \left(\frac{p4_j}{3}\right)$ should be minimized in the achievement function.

5 Results and Discussion

The complete FGP model described in Sect. 4 can be employed to build the schedule. In other words, the resulting schedule can be obtained by solving the following program.

$$Lex\min\ z = \left(\sum_{j=1}^{19} \left(\frac{n1_j}{2} + \frac{p1_j}{2} \right), \sum_{j=1}^{19} \left(\frac{n2_j}{1} + \frac{p2_j}{1} \right), \sum_{j=1}^{19} \left(\frac{p4_j}{3} \right) \right)$$

$s.t.$

$$\begin{aligned} &\text{constraints:} \\ &(7)\text{–}(26), \\ &(28), (30), (31), (33). \end{aligned} \tag{34}$$

In model (34), fuzzy goals 1–3 are set to priority levels 1–3 respectively. Indeed, model (34) is a multi-objective mixed integer programming problem that can be solved by optimisation packages such as LINGO (Schage 1999) with the use of the lexicographic methodology. It only requires a few minutes to solve the model to optimality for different calendar months. Table 1 shows a typical result comparison between a schedule developed by the head nurse (called manual schedule) and the resulting schedule of the developed model (called model schedule).

For a better comparison, some items of Table 1 are depicted in Figs. 2 and 3.

Figures 2 and 3 depict the unfairness of the distribution of night and day offs during weekends among nurses by manual schedule. It can be seen that the distribution obtained by the model is much preferred. Note that in Fig. 3a the nurses that are not present in the horizontal axis do not have any days off during the weekends.

In addition, the nurse scheduling of January 2008 in MW was generated by the proposed model. The schedule was welcomed by both nurses and hospital

Table 1 The results of comparison between manual and model schedules

Item	Manual schedule	Model schedule
Too many nurses work in some day shifts	Yes	No
Too many nurses work in some evening shifts	Yes	No
Too many nurses work in some night shifts	No	No
Does every nurse have more day shifts than evening shift?	No	Yes
Does every nurse have more day shifts than night shift?	No	Yes
Number of nurses with off shift each day	Uneven	Even
Distribution of day shifts among nurses	Unfair	Fair
Distribution of evening shifts among nurses	Unfair	Fair
Distribution of night shifts among nurses	Unfair	Fair
Distribution of public holidays and days off	Unfair	Fair
Distribution of off weekends	Unfair	Fair

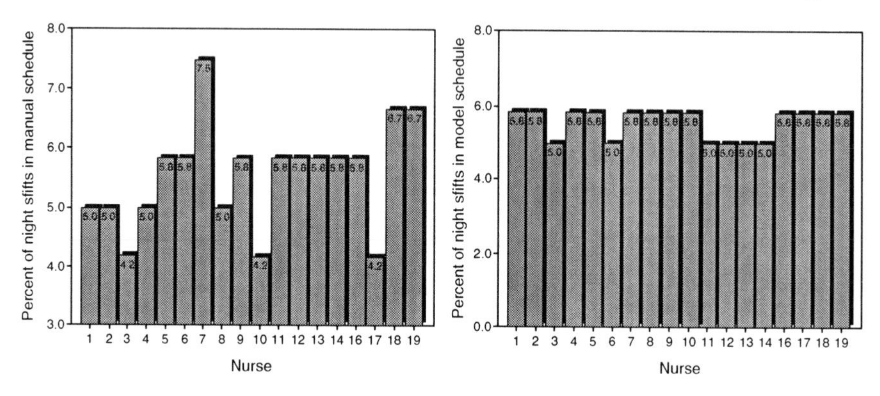

Fig. 2 Distribution of night shifts by manual and model schedules

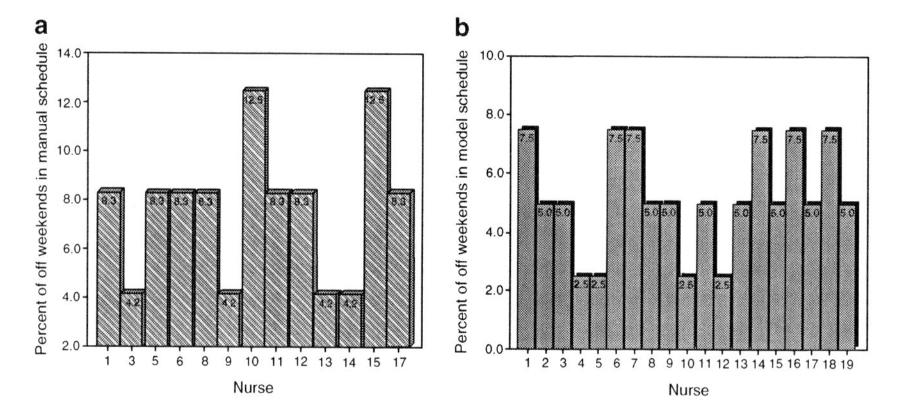

Fig. 3 Distribution of off weekends by manual and model schedules

managers. Moreover, the easy structure of the FGP model is such that it can be extended to include other objectives or preferences. Also, it has the potential of being implemented for similar or even different hospital wards elsewhere.

6 Concluding Remarks

This paper has developed a fuzzy goal programming model for obtaining nurse scheduling for a maternity ward in a hospital in Kerman, Iran. The monthly schedule developed by the proposed model takes into account both hospital's requirements and nurses' preferences. Amongst hospital objectives are continuous patient care and minimum number of nurses with appropriate nursing skills in each shift. Amongst the nurses' preferences are more day shifts than night shifts, more day shifts than evening shifts and at most three continuous night shifts. The objectives and nurses' preferences are divided into two major parts in the proposed model: hard constraints and fuzzy goals. Then a fuzzy goal programming based on the previous research of the authors is developed (Yaghoobi et al. 2008; Yaghoobi and Tamiz 2007). The resulting shifts are implemented and are much preferred to the current schedule as they are fairer shift distributions for the nurses as well as satisfying all the hospital's requirements.

The proposed model is easily applicable, takes a short time to solve, and can be extended to include other objectives and/or preferences as well as other hospital staff such as the Doctors.

At the time of writing this paper, the authors are experimenting with applying other variants of GP to the same problem in order to make a comprehensive study in comparing their performances.

As a possible extension to the problem it would be a good idea to study the actual demand on the number of different skilled nurses required during each shift. For this study the numbers are given by the hospital and are fixed for all the shifts.

References

Aickelin U, White P (2004) Building better nurse scheduling algorithms. Ann Oper Res 128: 159–177

Azaeiz MN, Al-Sharif SS (2005) A 0–1 goal programming model for nurse scheduling. Comput Oper Res 32:491–507

Bard JF, Purnomo HW (2005) Preference scheduling for nurses using column generation. Eur J Oper Res 164:510–534

Bester MJ, Nieuwoudt I, Van Vuuren JH (2007) Finding good nurse duty schedules: a case study. J Scheduling 10:387–405

Blake JT, Carter MW (2002) A goal programming approach to strategic resource allocation in acute care hospitals. Eur J Oper Res 140:541–561

Burke EK, de Causmaecker P, van den Berghe G, van Landeghem H (2004) The state of the art of nurse rostering. J Scheduling 7:441–499

Chanas S, Kuchta D (2002) Fuzzy goal programming- one notion, many meanings. Control Cybern 31:871–890

Chu SCK, Mpp H, Kky L, Hp L (2000) Nurse allocation models for maternal and child health services. J Oper Res Soc 51:1193–1204

Hannan EL (1981) On fuzzy goal programming. Decis Sci 12:522–531

Narasimhan R (1980) Goal programming in a fuzzy environment. Decis Sci 11:325–336

Romero C (2004) A general structure of achievement function for a goal programming model. Eur J Oper Res 153:675–686

Schage L (1999) LINGO Release 6.0. LINDO System Inc, IL, USA

Sitompul D, Randhawa SU (1990) Nurse scheduling models: a stat-of-the-art review. J Soc Health Syst 2:62–72

Topaloglu S, Ozkarahan I (2004) An implicit goal programming model for the tour scheduling problem considering the employee work preferences. Ann Oper Res 128:135–158

Yaghoobi MA, Jones DF, Tamiz M (2008) Weighted additive models for solving fuzzy goal programming problems. Asia Pac J Oper Res 25:715–733

Yaghoobi MA, Tamiz M (2007) A method for solving fuzzy goal programming problems based on MINMAX approach. Eur J Oper Res 177:1580–1590

Zadeh LA (1965) Fuzzy sets. Inform Contr 8:338–353